Encyclopedia of
Mathematics and Society

Mathematical Development and Concepts

Encyclopedia of Mathematics and Society

Mathematical Development and Concepts

Editors
Sarah J. Greenwald
and
Jill E. Thomley
Applachian State University

SALEM PRESS
A Division of EBSCO Publishing
Ipswich, Massachusetts

Cover Photo: © Jon Feingersh/Blend Images/Corbis

Copyright © 2013, by Salem Press, A Division of EBSCO Publishing, Inc.
All rights reserved. No part of this work may be used or reproduced in any manner whatsoever or transmitted in any form or by any means, electronic or mechanical, including photocopy, recording, or any information storage and retrieval system, without written permission from the copyright owner. For permissions requests, contact proprietarypublishing@ebscohost.com.

ISBN 978-1-4298-3750-7

Printed in the United States of America

Contents

List of Contributors	ix
Addition and Subtraction	1
Axiomatic Systems	3
Binomial Theorem	5
Conic Sections	7
Coordinate Geometry	10
Cubes and Cube Roots	12
Curves	15
Equations, Polar	18
Exponentials and Logarithms	19
Function Rate of Change	21
Functions	24
Graphs	27
Infinity	29
Limits and Continuity	31
Linear Concepts	33
Mathematical Certainty	35
Mathematical Modeling	37
Mathematician Defined	40
Mathematics, Applied	42
Mathematics, Defined	47
Mathematics, Elegant	49
Mathematics, Theoretical	52
Mathematics, Utility of	57

Matrices	59
Measurement, Systems of	61
Measurements, Area	64
Measurements, Length	66
Measurements, Volume	70
Measures of Center	71
Measuring Time	73
Multiplication and Division	75
Normal Distribution	79
Number and Operations	81
Number Theory	85
Numbers, Complex	87
Numbers, Rational and Irrational	90
Numbers, Real	92
Parallel Postulate	94
Perimeter and Circumference	96
Permutations and Combinations	98
Pi	101
Polygons	103
Polyhedra	105
Polynomials	107
Probability	109
Proof	113
Pythagorean Theorem	117
Randomness	119
Sample Surveys	121
Scatterplots	123
Sequences and Series	125
Similarity	128
Squares and Square Roots	129
Surfaces	131
Trigonometry	133
Units of Area	136
Units of Length	138

Units of Mass	139
Units of Volume	141
Vectors	142
Zero	144

Resource Guide	**147**
Index	**151**

List of Contributors

Hyungryul Baik
 Cornell University
Ron Barnes
 University of Houston, Downtown
Linda Becerra
 University of Houston, Downtown
Robert A. Beeler
 Eastern Tennessee State University
Matt Boelkins
 Grand Valley State University
Sarah Boslaugh
 Washington University School of Medicine
Murray R. Bremner
 University of Saskatchewan
David Brink
 University College Dublin, Ireland
Ka-Luen Cheung
 The Hong Kong Institute of Education
Dogan Comez
 North Dakota State University
Justin Corfield
 Geelong Grammar School
Beth Cory
 Sam Houston State University
Vesta Coufal
 Gonzaga University
Daniel Disegni
 Columbia University
Catherine C. Galley
 Independent Scholar
Lidia Gonzalez
 City University of New York
Rick Gorvett
 University of Illinois at Urbana-Champaign
Sarah J. Greenwald
 Appalachian State University
Simone Gyorfi
 O. Goga High School, Jibou, Romania
Gareth Hagger-Johnson
 The University of Leeds
Holly Hirst
 Appalachian State University
Calli A. Holaway
 University of Alabama
Jerry Johnson
 Western Washington University
Michael "Cap" Khoury
 University of Michigan, Ann Arbor
Carmen M. Latterell
 University of Minnesota, Duluth
Mariana Montiel
 Georgia State University
Ashwin Mudigonda
 Universal Robotics Inc.
Gregory Rhoads
 Appalachian State University
David C. Royster
 University of Kentucky
Carl R. Seaquist
 Texas Tech University
Abhijit Sen
 Suri Vidyasagar College
Padmanabhan Seshaiyer
 George Mason University
Shahriar Shahriari
 Pomona College
Kelli M. Slaten
 University of North Carolina, Wilmington

Tristan Tager
  Indiana University
Courtney K. Taylor
  North Greenville University
Jill E. Thomley
  Appalachian State University
Elena Toneva
  Eastern Washington University
Carlos J. Vilalta
  Center for Economic Research and Teaching

Jiri Wackerman
  *Institute for Frontier Areas of
  Psychology and Mental Health*
Sharon Whitton
  Hofstra University
Linda Reichwein Zientek
  Sam Houston State University

Addition and Subtraction

Category: History and Development of Curricular Concepts.
Fields of Study: Communication; Connections; Number and Operations; Representations.
Summary: Addition and subtraction are binary mathematical operations, each the inverse of the other, and are among the oldest mathematical concepts.

Addition can be thought of as a process of accumulation. For example, if a flock of 3 sheep is joined with a flock of 4 sheep, the combined flock will have 7 sheep. Thus, 7 is the sum that results from the addition of the numbers 3 and 4. This can be written as $3 + 4 = 7$ where the sign "+" is read "plus" and the sign "=" is read "equals." Both 3 and 4 are called addends. Addition is commutative; that is, the order of the addends is irrelevant to how the sum is formed. Subtraction finds the remainder after a quantity is diminished by a certain amount. If from a flock containing 5 sheep, 3 sheep are removed, then 2 sheep remain. In this example, 5 is the minuend, 3 is the subtrahend, and 2 is the remainder or difference. This can be written as $5 - 3 = 2$ where "−" is read "minus." Subtraction is not commutative and therefore the ordering of the minuend and subtrahend affects the result: $5 - 3 = 2$, but $3 - 5 = -2$.

The concept of addition can be extended to have meaning for fractions, negative numbers, real numbers, measurements, and other mathematical entities. The algorithms used for computing the sum or difference, some of which have been taught for millennia, ultimately depend on the representation used for the numbers. For example, the approach used for adding Roman numerals is different from that used to add Hindu-Arabic numbers. Computers perform subtraction using the same circuits they use for addition.

History and Development of Addition and Subtraction

Human beings' ability to add and subtract small whole numbers is probably innate. Some of the earliest descriptions of techniques for handling large numbers come from ancient China during the Warring States period (475–221 B.C.E.), when arithmetic operations were performed by manipulating rods on a flat surface that was partitioned by vertical and horizontal lines. The numbers were represented by a positional base-10 system. Some scholars believe that this system—after moving westward through India and the Islamic Empire—became the modern system of representing numbers.

The Greeks in the fifth century B.C.E., in addition to using a complex ciphered system for representing numbers, used a system that is very similar to Roman numerals. It is possible that the Greeks performed arithmetic operations by manipulating small stones on a large, flat surface partitioned by lines. A similar stone tablet was found on the island of Salamis in the 1800s and is believed to date from the fourth century B.C.E. The word "calculate" was derived from the Latin word for "little stone."

The Romans had arithmetic devices similar in appearance to the typical Chinese abacus. It is difficult to use modern paper-and-pencil techniques for adding and subtracting Roman numerals (with I as one, II as two, V as five, X as ten, L as fifty, C as one hundred, D as five hundred, M as one thousand)—but it worked well in its time, since it was devised for use with an abacus.

During the Middle Ages, counting boards were used to perform arithmetic. A counting board consisted of a series of actual or virtual horizontal lines that were labeled from the bottom by I, X, C, M, and so on. The system borrowed the symbols used for core numbers from the Roman system. The spaces between the lines were labeled starting from the bottom by V, L, and D. A number like MMDCCXXXVIIII (2739) would be represented by placing the appropriate number of counters on each line. The line labeled M would have 2 counters (for 2000, or two thousands). The space just below, labeled D, would have 1 counter (500, or one five-hundreds); the line labeled C, 2 counters (200, or two hundreds); the space labeled L, 0 counters; the line labeled X, 3 counters (for 30, or three tens); the line labeled V, 1 counter (5); and the line labeled I, 4 counters (4, or four ones). The total of all these numbers is 2739. Note that accountants used VIIII (denoting five plus four) to represent 9, whereas stonemasons used "IX"(denoting 10 less 1). To compute the sum MMDC-CXXXVIIII + MCLXI, a person would simply transcribe the numbers to the counting board and then combine the counters following rules of carrying to ensure that no more than 4 counters were on any line and 1 counter on any space. This representation was then easily transcribed back into Roman numerals.

Many early books on arithmetic claim that this method of performing arithmetic was especially preferred by women, who at times had the responsibility for keeping the books for small family businesses. Hindu-Arabic numerals and paper-and-pencil methods for performing arithmetic began to appear in Europe in the twelfth century and replaced Roman numerals and the counting board by the nineteenth century.

Two Methods for Subtracting by Hand

Two popular methods for handling "borrowing" that are taught today are shown below. The method shown in the figure below on the left is popular in Italy, England, and the United States, while the one on the right is popular in Spain, France, and parts of Latin America. The example is to compute 3047−1964. Starting with the method on the left, first begin with the rightmost column and subtract 4 from 7. Write the result, 3, below the 4. Moving one column to the left, try to subtract 6 from 4, which cannot be done without using negative numbers. The method is thus to attempt to "borrow" 1 from 0, which is the digit to the left of the 4. Again, this cannot be done without using negative numbers. Therefore, the method is to borrow 1 from 3, which is the digit to the left of the 0 resulting in crossing out the 3 and replacing it with a 2. Then the zero becomes a 10, and it in turn can be replaced by a 9 so the borrowed 1 can be placed in front of the 4 to make it 14. Now, one can subtract 6 from 14 to get 8, which is written below the 6. Moving left to the next column, one can subtract

9 from 9 to get a 0, which is written below the 9. Finally, 1 is subtracted from 2 to get a 1, which is written below.

$$\begin{array}{r} {}^{2}\cancel{3}\,{}^{9}\cancel{0}\,{}^{1}4\,7 \\ -1\,9\,6\,4 \\ \hline 1\,0\,8\,3 \end{array} \qquad \begin{array}{r} 3\,0\,4\,7 \\ -1\,{}_{1}9\,{}_{1}6\,4 \\ \hline 1\,0\,8\,3 \end{array}$$

To solve the problem using subtraction with carry, use the example on the right. The carrying numbers (the small 1s) affect the numbers on a diagonal, as shown in the example. The number 1 adds 10 to the integer in the top row and adds 1 to the integer in the bottom row. Starting from the rightmost column, 4 is subtracted from 7, resulting in 3, which is written below. Then, try to subtract 6 from 4, which cannot be done, so insert a small 1 to the left of the space between the 4 and the 6. This is interpreted to mean that the 4 has become 14. Subtract 6 from 14 and record the answer, 8, below. Move left to the next column containing 0 and 9. The small 1, written above and to the right of the 9, is added to the 9 to get 10. Attempt to subtract the 10 from the 0 above, which cannot be done. Instead, write a small 1 just to the left of the space between the 0 and 9, and interpret this to mean that the 0 has become a 10. Now, 10 minus 10 is 0, which is written below. Move left to the next column. The small 1, written above and to the right of the 1, is added to the 1 giving 2, which is subtracted from 3 resulting in 1, which is written below.

Adding and Subtracting on a Computer

At the most basic level, whole numbers are represented in a computer in base-two by a sequence of the binary states "Hi" and "Lo" interpreted as "1" and "0." The circuits that perform addition are implemented by sequences of logical gates. Typically a "1" in the leftmost bit indicates that the number is negative, with the remaining bits indicating the magnitude of the number. Subtraction can be performed by the same circuits that perform addition. Two popular approaches are designated as "one's complement" and "two's complement." "One's complement" can best be explained by performing subtraction in base-10 using "nine's complement." Assume a computation of 3047−1964. To find the "nine's complement" of 1964, subtract each digit from 9 to obtain 8035. This is added to 3047 resulting in 11,082. The leftmost 1 is viewed as a "carry" and brought around and added to the rightmost digit in an operation called "end-around carry" to obtain the final result: 1083.

Generalizing Addition and Subtraction

The sum of two fractions a/b and c/d is defined to be

$$\frac{ad + bc}{bd}.$$

The sum of irrational numbers (numbers that cannot be represented as fractions of whole numbers) can be approximated only by adding their approximating rationals. The exact sum of two irrational lengths, a and b, can be found exactly using geometry by first extending the segment representing a sufficiently on one end so that the length b can be marked off from that end with a compass.

Addition can be generalized to other mathematical objects, such as complex numbers and matrices. One of these objects, typically called the additive identity and denoted by "0," has the property such that if "a" is any object then the sum of 0 and a is a. The additive reciprocal of an object a is denoted by $-a$ and is defined to the object so that the sum $a + (-a)$ is 0. The difference $a - b$ is defined to be $a + (-b)$.

Further Reading

Flegg, G. *Numbers: Their History and Meaning.* New York: Schocken Books, 1983.

Karpinski, L. C. *The History of Arithmetic.* New York: Russell & Russell, 1965.

Pullan, J. M. *The History of the Abacus.* New York: F. A. Praeger, 1969.

Rafiquzzaman, M. *Fundamentals of Digital Logic and Microcomputer Design.* Hoboken, NJ: Wiley, 2005.

Yong, L. L., and A. T. Se. *Fleeting Footsteps.* Singapore: Word Scientific Publishers, 2004.

Carl R. Seaquist
Catherine C. Galley

Axiomatic Systems

Category: History and Development of Curricular Concepts.
Fields of Study: Communication; Connections; Geometry; Reasoning and Proof.
Summary: An axiom is a statement that is assumed to be true, and axiomatic systems have a rich and interesting mathematical history.

Axiomatic systems provide a deductive framework for mathematicians to combine related definitions and theorems that make mathematical knowledge systematic and structural. Mathematical theories including number systems, set theory, probability, algebra, and many others are built by using axiomatic systems.

Axiomatic Method and Axiomatic System in Mathematics

To build a deductive mathematical system, one needs to observe two intrinsic limitations in this process.

Limitation 1: Not every mathematical term can be defined. The reason can be seen by the following considerations: To define a term A, one needs a term B, and possibly some other terms. To define the term B, one needs another term, C, and so on. One may eventually come back to the term A; in which case the definition would be circular as there are a finite number of words. This means that A is used to define A, which is undesirable.

If the definitions are not to become circular, some terms are needed to start with. The solution is that there will be some terms that will not be defined. These will be called "undefined terms," and will be used to define all the other terms to be considered. One may think that it is strange that this solution can work. How can undefined terms give meaning? This puzzle is partially answered upon consideration of the next limitation.

Limitation 2: Not every mathematical statement can be deduced or proven. The reason is similar to the one in Limitation 1; some statements are needed to start a chain of deduction: if R, then S; if S, then T; if T, then U; and so on. To deal with this limitation, certain statements must be accepted without proof. These statements are called "axioms," and they are the statements that we used to deduce other statements. Actually, the axioms are often statements about the undefined terms. In other words, the axioms often tell us certain properties or restrictions of the undefined terms. Thus, the axioms help provide meaning to the undefined terms. Starting with undefined terms, axioms, and definitions, and by using deductive reasoning to establish important mathematics facts in the form of theorems, the mathematics system so obtained is said to be built by using the "axiomatic method." Such a system that consists of undefined terms, axioms A, definitions D, statements of the form If P then Q and proof of such statements is called an "axiomatic system." In an axiomatic system, one does not talk about the validity of *A* or *P*, one talks only about the validity of the proof based on *A* and *D*.

Historical Developments

Historically, Euclidean geometry was the best-known model of an axiomatic system. Around 300 B.C.E., Euclid wrote his 13-volume *Elements*, which contained an axiomatic treatment of geometry. It starts with 23 definitions; Euclid stated 10 axioms. The first five axioms are geometric assumptions, which he called postulates. The last five are more general, which he called common notions. There, Euclid did not use undefined terms.

The most important and fundamental property of an axiomatic system is "consistency" (it is impossible to deduce from these axioms a theorem that contradicts any axiom or previously proved theorem). The Euclidean geometry provides such a consistent axiomatic system. An individual axiom is "independent" if it cannot be logically deduced from the other axioms in the system. The entire set of axioms is said to be independent if each of its axioms is independent. Mathematicians prior to the nineteenth century doubted very much about the independence of the fifth postulate (the parallel postulate). They tried to deduce such a postulate by using the first four postulates. Despite considerable effort devoted to the task, no significant result could be obtained.

Euclid's *Elements* indeed became the most influential book on geometry, as well as the model of logical reasoning and axiomatic system, until the nineteenth century when two fundamental developments took place. First, it was realized that Euclid's logical system was not rigorous enough. A rigorous axiomatic treatment of Euclidean geometry was given by David Hilbert (1862–1943) in his 1899 book *Grundlagen der Geometrie* (*The Foundation of Geometry*). Here, Hilbert used the undefined terms of point, line, lie on, between, and congruent for the geometry system. Second, research results of C. F. Gauss (1777–1855), J. Bolyai (1802–1860), and N. I. Lobachevsky (1793–1856) asserted that the parallel postulate was actually an independent axiom. Non-Euclidean geometry could be developed by replacing the fifth postulate with another independent axiom. The lesson from Euclidean and non-Euclidean geometry is that both are valid axiomatic systems. When studying Euclidean or non-Euclidean geometry, no claims are made on the truth of the axioms about the physical world. One merely claims that if the axioms are valid, then the theorems deduced therein are also valid. Whether the logical system describes the real world is another question.

Current Issues

There are still many issues regarding the axiomatic systems. The set of axioms in an axiomatic system is "complete" if the axioms are sufficient in number to prove or disprove any statement that arises concerning our collection of undefined terms. To determine whether an axiomatic system is complete is by no means an easy question to answer. A great surprise was discovered by Kurt Gödel (1906–1978) in 1931. He proved that in a formal mathematics system that included the integers, there exist statements that are impossible to prove or disprove. This result is called Gödel's incompleteness theorem. Also, to determine whether a given proposition is an axiom has been a very important issue in computer science and is important when one tries to use a computer to do proofs. If the computer cannot recognize the axioms, the computer will also not be able to recognize whether a proof is valid.

Further Reading

Greenberg, Martin Jay. *Euclidean and Non-Euclidean Geometries: Development and History*. 3rd ed. New York: W. H. Freeman and Co, 1993.

Heath, Thomas L. *The Thirteen Books of Euclid's Elements*, Vol. 1–3. 2nd ed. New York: Dover Publications, 1956.

Meyer, Burnett. *An Introduction to Axiomatic Systems*. Boston: Prindle, Weber & Schmidt, 1974.

Venema, G. A. *The Foundations of Geometry*. New Jersey: Pearson Prentice Hall, 2006.

Wallace, E. C., and S. F. West. *Roads to Geometry*. 3rd ed. Upper Saddle River, NJ: Prentice Hall, 2003.

KA-LUEN CHEUNG

Binomial Theorem

Category: History and Development of Curricular Concepts.
Fields of Study: Algebra; Communication; Connections; Number and Operations.
Summary: The binomial theorem is the basis of Pascal's Triangle and is used to solve a variety of problems.

A binomial is an algebraic expression with two terms, like $x+y$. When binomials are multiplied together, they produce higher powers of the individual terms that are called "binomial coefficients." The binomial theorem states that for any real numbers x and y, and whole number n:

$$(x+y)^n = c_0 x^n y^0 + c_1 x^{n-1} y^1 + c_2 x^{n-2} y^2 + \cdots + c_n x^0 y^n$$

where c_k is the binomial coefficient

$$\frac{n!}{k!(n-k)!}$$

and $n!$ is the product of the numbers 1 through n. These coefficients are the entries in what is referred to as Pascal's Triangle, named for mathematician Blaise Pascal. Students typically encounter this theorem in middle school or high school algebra, and in high school or college calculus. It also has uses in other areas of mathematics, such as in combinatorics, where it helps in calculations for certain counting problems.

The binomial theorem is found across ages and cultures. It appears in the ancient world in the work of Greek mathematician Euclid of Alexandria. His formula was for the square ($n = 2$) of a binomial, but it was described geometrically rather than algebraically. There is also evidence that the Hindu scholar Aryabhata knew the theorem for cubes in the sixth century. At least as early as the eleventh century, Chinese mathematicians such as Jia Xian and later Zhu Shijie knew the binomial coefficients in the form of Pascal's Triangle. They used the binomial theorem to find square and cube roots, and evidence suggests they knew of the binomial theorem for large values of n. Around the fifteenth century, the binomial theorem and binomial coefficients to at least the seventh power were found in the writings of Islamic scholars including Omar Khayyam, Abu Bekr ibn Muhammad ibn al-Husayn Al-Karaji, Abu Ali al-Hasan ibn al-Haytham (Alhazen), and Ibn Yahya al-Maghribi al-Samaw'al.

In the sixteenth century, European mathematicians began using the binomial theorem and binomial coefficients. For example, mathematician Michael Stifel's 1544 work *Arithmetica integra* contained the binomial coefficients. Other contributors include François Viète, Blaise Pascal, James Gregory, Sir Isaac Newton, and Niels Abel. John Wallace's seventeenth century book *De Algebra Tractatus* is cited as the first published account of Newton's binomial work. While Pascal was not the first to study the binomial coefficients, he is credited with linking algebraic and combinatorial interpretations of the coefficients.

Pascal's Triangle is a triangular representation of the binomial coefficients, which may be attributed to him because of his 1653 work *Traité du Triangle Arithmétique* in which he compiled and expounded on much

Probability Theory

In probability theory, the binomial distribution uses binomial coefficients in the computation of probabilities. A binomial distribution is used to model a situation or process in which a series of independent trials occurs. Each trial may have only one of two possible outcomes, traditionally labeled "success" and "failure." In each trial, the chance of success or failure is constant, such as flipping a fair coin and getting a head, or rolling a fair die and getting a 6.

In this context, the binomial coefficient indicates the number of permutations there may be of a specific number of successes in a given number of trials; for example, the orderings of 2 heads and 8 tails in a series of 10 coin tosses. Mathematicians such as Jacob Bernoulli, Abraham de Moivre, Pierre de Laplace, Simeon Poisson, and Pascal worked on the binomial distribution and extensions, such as the limiting Poisson distribution. Mathematician and statistician Samuel Wilks, as well as others, developed the multinomial distribution to extend the binomial to cases with more than two possible outcomes on each trial.

of what was known about binomial coefficients. The related Pascal matrix is a symmetric, positive definite matrix with the Pascal triangle represented on its antidiagonals.

Generalizations and Extensions

James Gregory and Sir Isaac Newton generalized the binomial theorem to allow first fractional, and then real powers, which requires replacing the finite sum with an infinite series and extending the definition of the binomial coefficients. When generalizing to an infinite series, another issue that must be considered is convergence, which imposes restrictions on the numbers x and y for which the series converges to the binomial. Newton came to this generalization indirectly while trying to calculate areas under certain curves.

Another way to generalize the binomial theorem is to broaden the types of values that x and y can take. One such generalization allows x, y, and n to be complex numbers. The definition of the binomial coefficients has to be generalized to complex numbers, and certain restrictions on the variables are required for convergence of the resulting infinite series. Alternatively, one can allow x and y to be commuting elements of a Banach algebra, a normed algebra studied in such fields as complex analysis, real analysis, and functional analysis. Banach algebra is named for twentieth century mathematician Stefan Banach.

One more type of generalization considers not just the sum of two numbers x and y, but sums with more terms. Such a sum would be called a multinomial, and the multinomial coefficients would be appropriate generalizations of the binomial coefficients. Pascal's Pyramid or Pascal's Simplex are extensions of Pascal's Triangle for three or more dimensions.

Applications

The binomial theorem gives a quick way of expanding a power of the form $(x+y)^n$, making the formula useful for basic algebraic calculations. The binomial theorem, along with De Moivre's formula, can be used to prove the trigonometric double-angle identities, as well as more general formulas for $\cos(nx)$ and $\sin(nx)$.

The mathematical constant e also can be written as the infinite limit of
$$\left(1+\frac{1}{n}\right)^n.$$

Mathematical induction and the binomial theorem, or the multinomial theorem, can be used to prove what is known as "Fermat's little theorem," named for mathematician Pierre de Fermat. This result in number theory states that if p is a prime number and n is an integer not divisible by p, then $n^p - n$ is divisible by p. Fermat's little theorem is itself used in cryptography, providing an indirect application of the binomial theorem. One theorem in graph theory states that a graph with n vertices and adjacency matrix A is connected if and only if all the entries in the matrix $(1+A)^{n-1}$ are positive. This theorem is proved using the binomial theorem, generalized to certain matrices, and some basic graph theory results. Certain colorings of Pascal's Triangle produce fractal figures like Sierpinski's Triangle, named for mathematician Waclaw Sierpinski. In set theory, the regions of a Venn diagram for n distinct sets are in one-to-one correspondence with the binomial coefficients c_k for k ranging from 0 to n. Venn diagrams are named for John Venn.

Further Reading

"The Binomial Theorem." In Math. http://www.intmath.com/series-binomial-theorem/4-binomial-theorem.php.

Chauvenet, William. *Binomial Theorem and Logarithms.* Self-published: Biblio Bazaar, 2008.

Coolridge, J. L. "The Story of the Binomial Theorem." *American Mathematical Monthly* 56, no. 3 (March 1949).

Friedberg, Stephen H. "Applications of the Binomial Theorem." *International Journal of Mathematical Education in Science and Technology* 29, no. 3 (1998).

Fulton, C. M. "Classroom Notes: A Simple Proof of the Binomial Theorem." *American Mathematical Monthly* 59, no. 4 (1952).

Vesta Coufal

Conic Sections

Category: History and Development of Curricular Concepts.
Fields of Study: Algebra; Communication; Connections; Geometry.
Summary: Conic sections have many interesting mathematical properties and real-world applications.

Conic sections, or simply "conics," are the simplest plane curves other than straight lines. Students in the twenty-first century begin to study these curves in middle school. In coordinate geometry, they can be expressed as polynomials of degree 2 in two variables while straight lines are polynomials of degree 1 in two variables. Conic sections can further be divided into three types: ellipse, parabola, and hyperbola. Conics were named and systematically studied by Apollonius of Perga (262–190 B.C.E.). At that time, the study of conics was not merely to explore the intrinsic beauty of the curves but to develop useful tools necessary for applications to the solution of geometric problems. Today, the theory of conics has numerous applications in our daily lives including the designs of many machines, optical tools, telecommunication devices, and even the tracks of roller coasters.

Representations of Conics and Their Applications

One can generate a two-sheet circular cone by fixing a straight line as the axis of the cone in the space first. Choose a fixed point on it as the vertex of the cone. Rotating another straight line through the vertex that makes a fixed angle with the axis, we obtain the desired cone as the trace of the rotating line. Any straight line on the trace is called a "generating line" of the cone. Conic sections are obtained by intersecting the two-sheet cone with planes not passing through its vertex as shown in Figures 1A–C.

The three types of conic sections are generated according to the positions of the intersection.

Ellipse

When the intersecting plane cuts only one sheet of the cone and the intersection is a closed curve, an ellipse is created. A circle is obtained when the intersecting plane is perpendicular to the axis of the cone; an ellipse is obtained when the intersecting plane is not perpendicular to the axis of the cone. A circle, as such, can be considered as a particular case of an ellipse (Figure 1A). As illustrated in Figure 2, an ellipse is the collection of points in a plane that the sum of distances from two fixed points F_1 and F_2, the foci, to every point in the collection is constant.

On the coordinate plane, if the foci are located on the x-axis at the points $(-c, 0)$ and $(c, 0)$ and the constant distance between F_1 and F_2 is $2a$, the equation of an ellipse can be derived as

$$\frac{x^2}{a^2} + \frac{y^2}{b^2} = 1 \text{ with } b^2 = a^2 - c^2 < a^2.$$

Parabola

When the intersecting plane cuts only one sheet of the cone and is parallel to exactly one generating line of the cone, the intersection is a non-closed curve—a parabola (Figure 1B). A parabola is the collection of points in a plane that are equidistant from a fixed point F (called "focus") and a fixed line (called "directrix"). The graph of the parabola is illustrated in Figure 3A. The graph is symmetric with respect to the line through the focus and perpendicular to the directrix. This line of symmetry is called the "axis" of the parabola. The intersection of the graph with the axis is called the "vertex" of the parabola. On the coordinate plane, if the vertex is located at the origin O, and the focus at the point $(0, p)$, then its directrix will be on the line $y = -p$ (Figure 3B), and the equation of the parabola can be derived as $x^2 = 4py$.

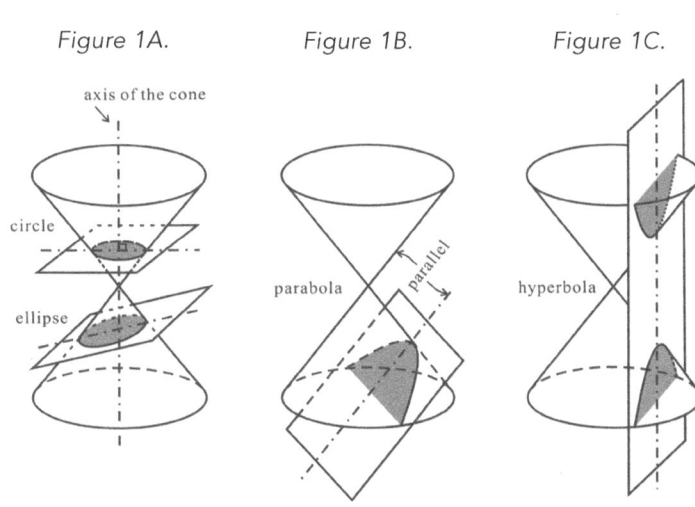

Figure 1A.   Figure 1B.   Figure 1C.

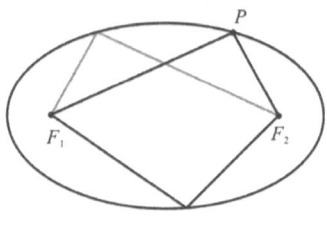

Figure 2.

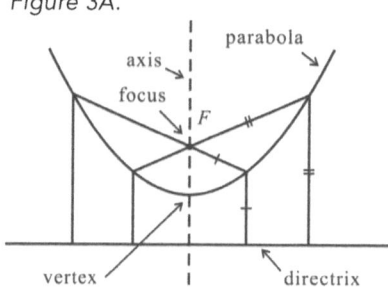

Figure 3A.

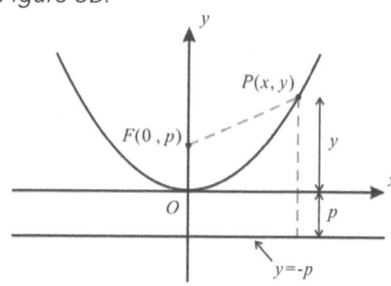

Figure 3B.

Hyperbola

When the intersecting plane meets both sheets of the cone, the intersection is a hyperbola, which consists of two identical non-closed parts, each located in one of the two sheets of the cone (Figure 1C). A hyperbola is the collection of all points in a plane that the difference of distances from two fixed points F_1 and F_2, the foci, to every point in the collection is constant. The graph of a hyperbola is drawn as shown in Figure 4A.

On the coordinate plane, if the foci $(-c, 0)$ $(c, 0)$ are located on the x-axis and the differences of distance is $\pm 2a$, then the equation of the hyperbola can be derived as

$$\frac{x^2}{a^2} - \frac{y^2}{b^2} = 1 \text{ with } c^2 = a^2 + b^2 \text{ (See Figure 4B)}$$

A Brief History of Conic Sections

Between 460 B.C.E. and 420 B.C.E., three famous geometry problems were posed by the ancient Greeks. These problems were (1) the trisection of an angle, (2) the squaring of the circle, and (3) the duplication of the cube. The last problem merely asks that given any cube of side length a, can one construct another cube with exactly twice the volume, $2a^3$. Hippocrates of Chios (circa 470–410 B.C.E.) had the idea of reducing that problem by finding two quantities x and y such that

$$\frac{a}{x} = \frac{x}{y} = \frac{y}{2a}.$$

Then, $x^2 = ay$, $y^2 = ax$, and $xy = 2a^2$.

As such, x is the required solution for the problem. This solution is equivalent to solving simultaneously any two of the three equations ($x^2 = ay$, $y^2 = 2ax$, and $xy = 2a^2$) that represent parabolas in the first two and a hyperbola in the third. However, no explicit construction of the conic sections was given. Menaechmus (380–320 B.C.E.) is believed to be the first mathematician to work with conic sections systematically, which is theorized to have arisen because of curves traced out by sundials. At his time, the conic sections were formed by cutting a right circular cone with a plane perpendicular to a side.

The sections were named according to whether the vertex angle was right, acute, or obtuse (Figure 5). Menaechmus constructed conic sections that satisfied

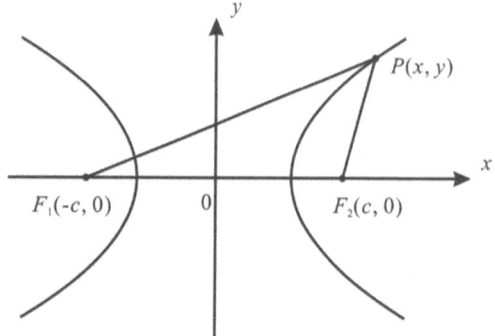

Figure 4A.

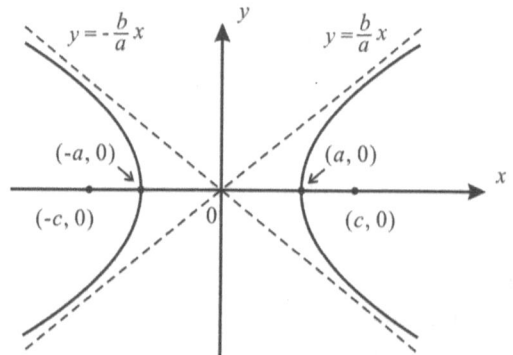

Figure 4B.

Figure 5.

  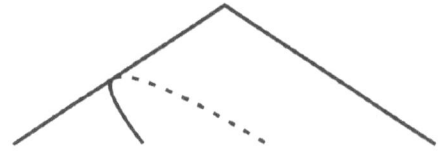

the required algebraic properties suggested by Hippocrates and thus obtained the points of intersection of these conic sections that would lead to the solution of the problem of the duplication of the cube.

The breakthrough in the study of conics by the ancient Greeks was attributed to Apollonius of Perga. His eight-volume masterpiece *Conic Sections* greatly extended the existing knowledge at the time (one of the eight books has been lost to history). Apollonius' major contribution was to treat the conic sections as plane curves and use their intrinsic properties to characterize them. This method allowed conic sections to be analyzed in great detail by the ancient Greeks.

Abu Ali al-Hasan ibn al-Haytham studied optics using conic sections in the tenth and eleventh centuries. Omar Al-Khayyami (Omar Khayyam) authored *Treatise on Demonstration of Problems of Algebra* in the eleventh century. This work showed that all cubic equations could be classified using geometric solutions that involve conic sections. Later, in the seventeenth century, Gerard Desargues (1591–1662) and Blaise Pascal (1623–1662) connected the study of conic sections to developments from projective geometry. At the same time, René Descartes (1596–1650) and Pierre de Fermat (1601–1665) also connected it with the developments from coordinate geometry. Eventually, problems of conics in geometry could be reduced to problems in algebra.

Johan Kepler (1571–1630) revolutionized astronomy by introducing the notion of elliptical orbits. According to Isaac Newton's later law of universal gravitation, the orbits of two massive objects that interact are conic sections.

If they are bound together, they will both trace out ellipses; if they move apart, they will both follow parabolic or hyperbolic trajectories.

The Applications of Conic Sections

Besides applications in astronomy, conics have many other applications.

In an ellipse, any light or radiation that begins at one focus will be reflected to the other focus (Figure 6). This property can be used in theater designs. In an elliptical theater, the speech from one focus can be heard clearly across the theater at the other focus by the audience. It can also be applied in lithotripsy, a medical procedure for treating kidney stones. The patient is placed in an elliptical tank of water, with the kidney stone fixed at one focus. High-energy shock waves emitted at the other focus can be directed to pulverize the stone. Also, elliptical gears can be used for many machine tools.

In a parabola, parallel light beams will converge to its focus (see Figure 7). Parabolic mirrors are used to converge light beams or heat radiations, and parabolic microphones are used to perform a similar function with sound waves.

In reverse, if a light source is placed at the focus of a parabolic mirror, the light will be reflected in rays parallel to said axis. This property is used in the design of car headlights and

Figure 6.

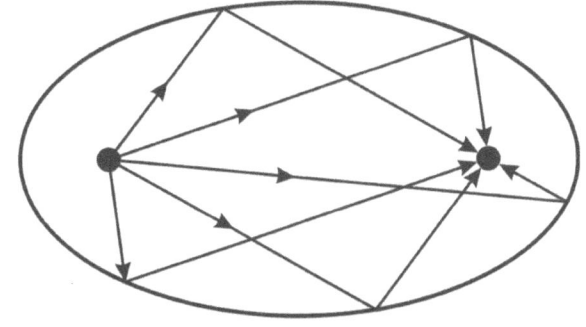

Figure 7.

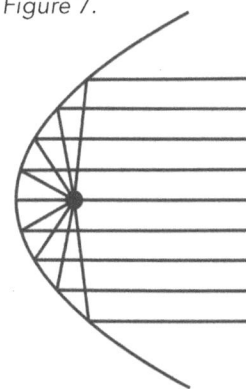

in spotlights because it aids in concentrating the parallel light beam. Hyperbolas are used in a navigation system known as Long Range Navigation (LORAN). Hyperbolic—as well as parabolic—mirrors and lenses are also used in systems of telescopes.

Further Reading

Akopyan, A. V., and A. A. Zaslavsky. *Geometry of Conics.* Providence, RI: American Mathematical Society, 2007.

Courant, R., and H. Robbins. *What Is Mathematics?* New York: Oxford University Press, 1996.

Downs, J. W. *Practical Conic Sections: The Geometric Properties of Ellipses, Parabolas and Hyperbolas.* Mineola, NY: Dover, 2003.

Kendig, K. *Conics (Dolciani Mathematical Expositions).* Washington, DC: The Mathematical Association of America, 2005.

Kline, M. *Mathematical Thought From Ancient to Modern Times.* New York: Oxford University Press, 1972.

Suzuki, Jeff. *A History of Mathematics.* Upper Saddle River, NJ: Prentice Hall, 2002.

Ka-Luen Cheung

Coordinate Geometry

Category: History and Development of Curricular Concepts.
Fields of Study: Algebra; Communication; Connections; Geometry.
Summary: The development of coordinate geometry revolutionized mathematics, has a wide variety of applications, and is now widely used in many areas of mathematics.

The discovery that plane geometric configurations could be entirely described by real number pairs and two-variable equations revolutionized geometry and many other important fields of mathematics that emerged later, including real analysis, vectors, calculus, linear algebra, and matrix theory. Also referred to as "analytic geometry" or "Cartesian geometry," named for the great philosopher and mathematician René Descartes, the subject of coordinate geometry is the study of geometry using the Cartesian coordinate system with algebraic operations. In twenty-first century classrooms, children in primary school begin to examine coordinate systems and create plots on graph paper.

The level of sophistication of knowledge builds through high school and college through the use of various coordinate systems including Cartesian, polar, and spherical systems and by representations in two- and three-dimensional geometry. Some calculus courses are titled "Calculus and Analytic Geometry." Various coordinate system standards are in use in physics or mathematics, for surveyors, or at the state or company level. High school and college students learn to convert between some of these representations. Coordinate geometry has many applications and is used in every conceivable area of mathematics, science, and engineering to calculate precise locations and boundaries, distances and bearings from reference points, and to define graphs and curves using a point location, radius, and arc-lengths.

The fundamental building block of coordinate geometry is the Cartesian coordinate system, which includes an infinite collection of points on a plane determined by an ordered pair of numerical coordinates (x, y). The x-coordinate (called "abscissa") represents the horizontal position, and the y-coordinate (called "ordinate") represents the vertical position. These positions can be expressed as signed distances from the origin $(0, 0)$, a point that is at the intersection of two perpendicular reference lines called the "coordinate axis" (see Figure 1).

Once points are determined by ordered pairs (x, y) on the coordinate plane, one can then obtain analytic

Figure 1.

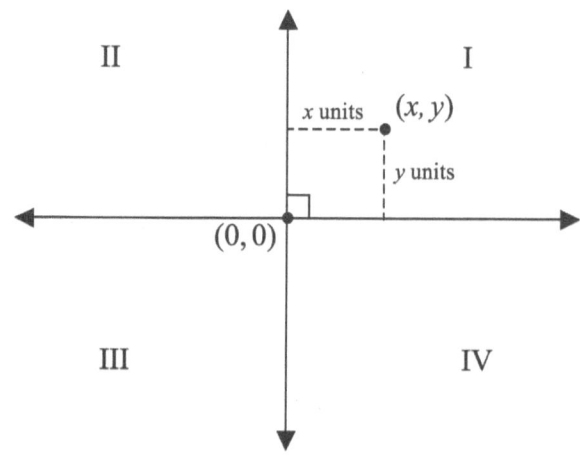

formulas for various geometric quantities on the plane. For example, an application of the Pythagorean theorem then yields the distance between any two points (x_1, y_1) and (x_2, y_2) given by

$$d = \sqrt{(x_2 - x_1)^2 + (y_2 - y_1)^2}.$$

Early Variations

Coordinate-like types of systems arose in cartography well before Descartes. Maps with grids date back to ancient times, including those by Dicaearchus of Messana and Eratosthenes of Cyrene. Claudius Ptolemy attempted to create coordinates of well-known places in the world, essentially their latitude and longitude, from spherical projections, although the astronomical and mathematical methods to accurately calculate these would not be completely developed until much later. Islamic Mathematicians in the medieval Islamic world, such as Abu Arrayhan Muhammad ibn Ahmad al-Biruni, who compared the work of Ptolemy and Abu Ja'far Muhammad ibn Musa Al-Khwarizmi, provided coordinates for more than 600 geographical locations. Al-Biruni also used rectangular coordinates to represent three-dimensional space as well as ideas that some consider as a precursor to polar coordinates. In the twenty-first century, the global positioning system calculates the coordinates of a user from a system of satellites.

Other aspects of coordinate geometry can also be found in various early contexts. Some have noted that the mathematical work of ancient Greek mathematician Menaechmus could be interpreted as one that used coordinates. However, there was no algebra in ancient Greece, and others have highlighted the challenge that mathematics historians face in judging historical works. Coordinate geometry is a natural leap for the historians but probably not for Menaechmus, critics assert. Graphing techniques were developed in the fourteenth century in publications of Nicole d'Oresme and a work titled *De latitudinibus formarum* (The Latitudes of Forms), which some attribute to d'Oresme. Others assert that this attribution is an error and that the author is unknown. These works may have influenced coordinate geometers.

Transformations of coordinate-like systems developed along with perspective drawing techniques of curves and shapes, like in the works of Leone Battista Alberti and Piero della Francesca. Polar coordinates were motivated through the work of mathematicians such as Bonaventura Cavalieri on spiral curves like the Archimedean spiral, named for Archimedes of Syracuse.

Development

Descartes and Pierre de Fermat are both credited with independently introducing coordinate geometry. They each introduced a type of single-axis system or ordinate geometry. Distances could be measured at a fixed angle to the reference line. In Fermat's work, curves are generated as loci rather than by plotting points. Historian of science Michael Sean Mahoney noted: "There is connected with the system an intuitive sense of motion or flow wholly in keeping with the intuition which underlies the notion of an algebraic variable." Descartes' published work on coordinate geometry dates to 1637 in the appendix (*La Géométrie*) of a short book entitled *Discourse on the Method*. Descartes defined the five algebraic operations of addition, subtraction, multiplication, division, and extraction of square roots as geometric constructions on line segments and showed how these operations could be performed in the Euclidean plane by straightedge-and-compass constructions. He also developed geometric techniques for solving polynomial equations by intersecting curves, such as conic sections, with each other or with lines to obtain solutions algebraically. Coordinate geometry helps to classify conic sections, which are curves corresponding to the general quadratic equation

$$ax^2 + bxy + cy^2 + dx + ey + f = 0$$

where a, b, c, d, e, and f are constants and a, b, and c are not all zero. Coordinate geometry became useful in a wide variety of mathematical and physical situations. Sir Isaac Newton and others investigated various coordinate systems as well as how to convert between them. In the nineteenth century, Christof Gudermann investigated the sphere, and Julius Plücker published numerous volumes on analytic geometry.

Variations

In situations where there is no obvious origin or reference axes, mathematicians developed local coordinates or coordinate-free approaches. For instance, the Frenet–Serret frame is named for Jean Frédéric Frenet and Joseph Serret. It is a type of coordinate axis system for a curve in three-dimensional space and represents the twists and turns of a curve as three vectors that move along the curve. Jean-Gaston Darboux explored the analog for a surface.

Another example is "isothermal coordinates" on surfaces in the work of mathematicians, like Carl Friedrich Gauss. Engineer, mathematician, and physicist Gabriel Lamé is noted as the first to use the term in his 1833 work on heat transfer. August Möbius introduced barycentric coordinates, which utilizes notions related to the center of mass and the centroid of a triangle, and these coordinates can be found in computer graphics. Möbius' work used both the position and magnitude.

Other mathematicians developed similar systems, including vectors, which allowed for compact notation. Hermann Grassmann and William Hamilton created the algebra of vectors. The development of vectors was especially useful when extending the geometry or physics to higher dimensions. A point (x, y) in the plane can also be represented by a vector as $r = x\hat{\imath} + y\hat{\jmath}$ where $\hat{\imath}$ and $\hat{\jmath}$ are unit vectors. Mathematicians including Jean-Victor Poncelet and Michel Chasles developed synthetic projective geometry, which focused on axioms instead of coordinates. Gregorio Ricci-Curbastro and Tullio Levi-Civita explored a coordinate-independent calculus, which led to the development of tensor analysis that later became important in general relativity. Bernhard Riemann's work on geodesics and Riemannian geometry led to geodesic coordinates, which also became important in relativity.

Education

Coordinate geometry took on an increased prominence in schools in the nineteenth and twentieth centuries. One reason was the development and curricular use of graph paper. A patent for printed graph paper dates back to Dr. Buston in the late eighteenth century. Graph paper makes it easier to plot points and create curves, and it was found to be useful in surveying and civil engineering projects.

Mathematicians in the nineteenth century, like E. H. Moore, advocated the use of paper with "squared lines" in algebra classes. Coordinate geometry topics were also included in algebra textbooks and in textbooks devoted to the subject.

One notable textbook was published by Scottish mathematician Robert J. T. Bell in 1910. His treatise on coordinate geometry in three dimensions became a very successful textbook on the subject and was translated into numerous languages.

Further Reading

Boyer, Carl B. *History of Analytic Geometry*. New York: Dover Books on Mathematics, 2004.

Mahoney, Michael Sean. *The Mathematical Career of Pierre de Fermat, 1601–1665*. 2nd ed. Princeton, NJ: Princeton University Press, 1994.

Sasaki, C. *Descartes's Mathematical Thought*. New York: Springer, 2010.

Padmanabhan Seshaiyer

Cubes and Cube Roots

Category: History and Development of Curricular Concepts.
Fields of Study: Algebra; Communication; Connections; Geometry.
Summary: Cubes and cube roots have been the subject of classical problems in mathematics, some of which were not solved for centuries.

Cubes and cube roots of numbers have played an important part in the development of mathematics. Middle school students are taught cubes and cube roots in order to solve equations and to calculate volumes of solids. In calculus, the cube root function is a common example of a function that is continuous everywhere but has an infinite derivative at one of its points. In addition, the cube function is an example of a function that is strictly increasing everywhere but has a point where the derivative is zero. The cube rule relates the percentage of popular vote in an election with the expected percentage of seats won in a two-party election. The power needed to overcome wind resistance is directly proportional to the cube of the wind speed. One model shows the heart rate in mammals is inversely proportional to the cube root of the weight of the animal. People in many different cultures have studied cubes and cube roots, and numerous interesting stories are found in its history. These simple objects have also generated many new ideas and new fields of mathematics.

Definition

To cube a particular number x, multiply it by itself 3 times—this is denoted x^3. If x is a number such that $x^3 = y$ for some other number y, then x is a cube root of y, written as $x = \sqrt[3]{y}$. Since $(-5)^3 = -5 \times -5 \times -5 = -125$, -5 is the cube root of -125, and the notation is

$$\sqrt[3]{-125} = -5.$$

The cube of any real number is unique; however, every real number has exactly one cube root that is a real number and two cube roots that are complex numbers.

Early History

As with squares, the earliest uses of cubes of numbers involved common geometric objects, specifically the cube, which is a three-dimensional object with six sides, all of which are congruent squares. The volume of a cube is the cube of the length of one of its sides. The volume of a sphere is directly proportional to the cube of its radius. One of the classical problems in Greek mathematics was the problem called "Duplication of the Cube." The problem was to find the length of an edge of a cube that has double the volume of a given cube using the tools of the time, the ruler and compass. It is now known that if x is the length of the side of the given cube, then $\sqrt[3]{2} \times x$ is the length of the cube with twice the volume. One possible origin of this problem is that, in 430 B.C.E., it was proclaimed through the oracle at Delos that the cubical altar to Apollo was to be doubled in volume in order to alleviate a plague that had befallen the people. Another possibility is that the Pythagoreans successfully doubled the square and doubling the cube was a natural extension. In any event, many great mathematicians throughout history worked on this problem, and, in the nineteenth century, it was proven that a solution was impossible.

Cube roots can be exact numbers if the cube root is an integer or a fraction. However, the cube root of most numbers is irrational (it has a infinite non-repeating decimal expansion) and its value can only be approximated. The easiest method to approximate the real cube root of a real number is to raise the number to the 1/3 power on a calculator. Obviously, the calculator is a recent invention, and other methods have been developed for approximating a cube root. Some of the earliest known methods are found in the Chinese text *Nine Chapters on the Mathematical Art* (c. first century C.E.) and in the book *Aryabhatiya* by the Indian mathematician Aryabhata (b. 476 C.E.). Both methods use the formula $(a + b)^3 = a^3 + 3a^2b + 3ab^2 + b^3$ repeatedly to generate the successive digits of the cube root. Approximations to cube roots can also be computed with the Chinese abacus, the *suànpán*, which dates to 200 B.C.E. In many cases, scribes would create tables of cube roots, which people would use to look up values for use. Barlow's Tables, named for mathematician Peter Barlow, who originally published in 1814, give the value of cubes and cube roots to nine decimal places and are still in print in the twenty-first century. Recreational mathematicians have found it fun to devise ways to compute cubes and approximations to cube roots in their head without outside assistance.

Cubic Equations

Cubic equations are equations that involve positive integer powers of x where the highest power is 3. Mathematicians have been trying to solve these equations from the earliest times. The Babylonian text *BM 85200* (c. 2000 B.C.E.) contains many problems that compute the volume of an excavated rectangular cellar by setting up and solving a cubic equation. Another Babylonian tablet contains, among other things, a table of integers and the sum of each integer's square and cube, and it was presumably used to solve cubic equations.

Archimedes of Syracuse (c. third century B.C.E.) considered the problem of passing a plane through a sphere such that the volumes of the two pieces had a certain ratio. This problem gives rise to what would be a cubic equation. A manuscript, thought to have been written by Archimedes, was found centuries later that gave a detailed solution to the problem that involved finding the intersection of a parabola with a hyperbola. Omar Khayyam (c. eleventh century C.E.) was the first to find a positive root of every cubic equation having one. Before this time, numbers were thought of as specific quantities of objects, so very little was done with negative numbers—and certainly not complex numbers. As with Archimedes, Khayyam's solutions involved intersecting conic sections.

By the fifteenth and sixteenth centuries, negative numbers and zero were accepted, and many of the Greek mathematical texts were translated to Latin. The field of algebra had been developed, and people could study equations as expressions with variables that can be manipulated (as is done in the twenty-first century).

As the solution of the general quadratic equation had been discovered, Italian mathematicians focused their attention to the solution of the general cubic equation $ax^3 + bx^2 + cx + d = 0$. During this period, academic reputations and employment were based on public problem-solving challenges, and discoveries were kept secret so they could be used to win one of these challenges.

The solution of certain cubic equations provided the backdrop to one of the more entertaining chapters in the history of mathematics. On his deathbed in 1526, Italian mathematician Scipione del Ferro told one of his students, Antonio Maria Fior, how to solve a specific type of cubic equation. Nine years later, Fior submitted 30 cubic equations of this type to mathematician Niccolo Tartaglia in a public challenge. During the contest, Tartaglia himself discovered the solution and won the contest. After hearing of the contest, Girolamo Cardano contacted Tartaglia to inquire about his method. Tartaglia told him his solution, only after Cardano agreed to keep it secret as Tartaglia indicated he was going to publish it (thinking he was the first to discover it). Years later, Cardano found out that del Ferro actually discovered the formula and published it as del Ferro's method in addition to solutions to the cubic equation in all cases that he and his assistant, Lodovico Ferrari, discovered. Tartaglia was extremely angry and felt Cardano had broken his promise. In the twenty-first century, the formula for the solution of the cubic is known as the Cardan(o)–Tartaglia formula.

Uses and Applications

The cubic equation also played an essential role in the formulation of complex numbers. In his 1572 text, *Algebra*, Rafael Bombelli considered the equation $x^3 = 15x + 4$. Applying the formula of Tartaglia and Cardano, one obtains a solution

$$x = \sqrt[3]{2 + \sqrt{-121}} + \sqrt[3]{2 - \sqrt{-121}} \ .$$

However, Bombelli knew the solution was actually 4 and that, somehow, the square root of −121 could be manipulated in a way to reduce this expression to 4. He developed an algebra for working with these roots of negative numbers (thought to be of no use to earlier mathematicians), and complex numbers and the field of complex analysis was born. With complex numbers, one can show that any cubic equation has exactly three solutions, two of which must be complex and one real.

In 1670, it was discovered that French mathematician Pierre De Fermat claimed that for all natural numbers $n > 2$ there are no nontrivial solutions of positive integers a, b, c such that $a^n + b^n = c^n$. Andrew Wiles proved this theorem in 1994 using objects called "elliptic curves." Elliptic curves are defined by a cubic equation of the form $y^2 = x^3 + ax + b$, whose graph has no cusps or self-intersections. These curves are studied in the twenty-first century and used in both number theory and cryptography (the study of coding information). Even though Fermat's equation has no positive integer solutions for $n = 3$, other problems involving sums of cubes have been studied.

In 1770, Edward Waring proposed the following question: for every positive natural number k, does there exist a natural number s such that every natural number N can be written as the sum of at most s numbers which are kth powers? If $k = 3$, the question becomes: can every positive number be written as a sum of at most s cubes? Some examples are $5 = 1^3 + 1^3 + 1^3 + 1^3 + 1^3$ and $23 = 2^3 + 2^3 + 1^3 + 1^3 + 1^3 + 1^3 + 1^3 + 1^3 + 1^3$.

As 23 shows, one requires at least 9 cubes. In 1909, David Hilbert proved that 9 is the maximum number of cubes that are required for any positive natural number. The Waring-Goldbach problem asks a similar question, except it requires at most s cubes of prime numbers. Some progress has been made, but this question remains unsolved as of 2010.

One of the more interesting recent mathematicians is Srinivasa Ramanujan from India (1887–1920). He was mostly self-educated and was able to prove theorems in number theory that shocked one of the eminent mathematicians of the time, G. H. Hardy. Once when Hardy visited Ramanujan, he mentioned he arrived in a cab numbered 1729, which did not seem very interesting. Ramanujan responded that 1729 is a very interesting number in that it is the smallest positive integer that can be represented by a sum of two cubes in two different ways, $1729 = 1^3 + 12^3 = 9^3 + 10^3$, which is correct. The taxicab numbers are generalizations of this idea. The nth taxicab number, denoted $Ta(n)$, is the smallest positive integer that can be written as two different cubes in n different ways. By Ramanujan's comment, $Ta(2) = 1729$. It is also true that $Ta(1) = 2$, since $2 = 1^3 + 1^3$ and $Ta(3) = 87,539,319$, since

$$87{,}539{,}319 = 167^3 + 436^3 = 228^3 + 423^3$$
$$= 255^3 + 414^3.$$

The first 6 taxicab numbers are known, but $Ta(7)$ and beyond are all unknown as of 2010.

Further Reading
Burton D. *A History of Mathematics: An Introduction.* 7th ed. New York: McGraw-Hill, 2011.

Dunham, William. *Journey Through Genius*, New York: Penguin Books, 1991.

Katz, Victor. "The Roots of Complex Numbers." *Math Horizons* 3 (1995).

Washington, Lawrence. *Elliptic Curves: Number Theory and Cryptography*. 2nd ed. Boca Raton, FL: CRC Press, 2008.

Gregory Rhoads

Curves

Category: History and Development of Curricular Concepts.
Field of Study: Algebra; Calculus; Communication; Connections; Geometry.
Summary: Curves have many different definitions and applications in various fields of mathematics.

Intuitively, a curve might be thought of as a path, like that of a curveball. A curve is viewed and defined in several ways depending on the branch of mathematics. A curve can be defined as the one-dimensional continuous trajectory of an object in space moving in time, the intersection of two surfaces in space, the image of the unit interval under a continuous function, or the graph of a solution of a polynomial equation. Each of these approaches captures the intuitive idea of a curve in their respective domains; the first is more physical, the second is geometric, the third is topological, and the last is an algebraic view of a curve. Curves can be used to create figures, model paths of motion, or express relationships between variables.

There are many types of curves that are the focus of classroom investigations, including yield curves, which are important to investors, and the normal distribution or bell-shaped curve. Felix Klein is noted to have said, "Everyone knows what a curve is, until he has studied enough mathematics to become confused through the countless number of possible exceptions." In education, a "learning curve" is a phrase that is meant to informally capture the notion of the change in knowledge over time. Algebraic and geometric curves are also important in school. Children study lines and circles in primary and middle school. They investigate their lengths and areas. By high school and college, they learn about parametric equations of curves and the area under a curve. In order to enrich classroom learning, mathematicians and mathematics historians created the National Curve Bank Web site.

Early History of the Study of Curves
The Greeks initiated the study of curves and discovered numerous interesting curves. Apollonius of Perga studied conic sections as the intersections of a plane and a cone by changing the angle of intersection. Diocles of Carystus invented the cissoid curve and used it in his attempts to solve the problem of doubling the cube. Nicomedes invented the conchoid curve and used it in his attempts to solve the problems of doubling the cube and trisecting an angle. Some have noted that aspects in the design of the columns of the Parthenon may resemble a conchoid of Nicomedes, although others present different curves as the model. Canon of Samos invented the spiral that was eventually called the "Archimedean spiral."

This curve was utilized by Archimedes of Syracuse as a method to attempt to trisect an angle and square the circle. The Greek view of curves was geometric, since Greek mathematics was, essentially, geometry-centered. Hence, their study of curves usually was through some elaborate and often ingenious methods of construction. Besides the lack of analytical tools, their insistence of having concrete or mechanical methods of construction, and—more importantly—their attempts to solve some important problems of antiquity that later were shown to be unsolvable by ruler and compass constructions are some of the factors contributing to the Greek concept of curves.

Mathematicians, philosophers, and others introduced and investigated the geometry of many interesting curves long after the ancient Greeks. For example, Nicholas of Cusa lived in the fifteenth century. He is noted as the first of many to explore the cycloid, which was eventually known as the path of a point on a wheel as the wheel rolls along a straight line.

Developments Since the Seventeenth Century

With the introduction of analytic geometry in the seventeenth century, the theory of curves received a new impetus—expressing curves by equations would make their study much easier compared to doing it via elaborate geometrical constructions. Analytic geometry enabled mathematicians to focus on the intrinsic features of curves; discover and investigate new curves; study curves in a more systematic way, leading to their classification into algebraic versus transcendental categories; and apply the results to various physical problems, such as the long-standing problem of determining the orbits of planets or solving the problem of a hanging chain, which was posed by Jacob Bernoulli.

Gottfried Leibniz, Christiaan Huygens, and Jacob Bernoulli's brother Johann Bernoulli responded to the elder Bernoulli's challenge with the equation of the catenary. In the eighteenth century, Guido Grandi investigated rhodonea curves that resemble roses and what was later to be known as the Witch of Agnesi, named because of a mistranslation of the example in Maria Agnesi's famous calculus textbook.

Beginning with the seventeenth century, smooth curves have been an intense subject of investigation leading to determination of various features. Smooth curves, like lines, circles, parabolas, spirals, and helices, possess properties that make them amenable to numerous applications besides lacking any jagged behavior. For example, younger students learn that a straight line is the shortest path between two points in the plane, and mathematicians in the seventeenth century wondered about an analog for surfaces. A geodesic curve is

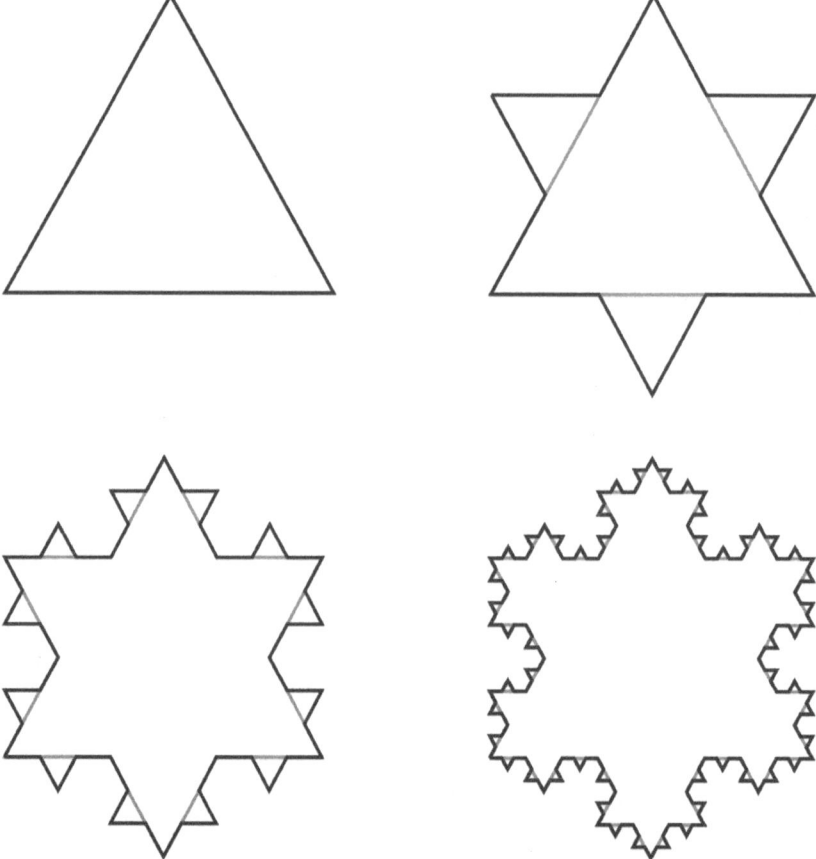

Curves can be extremely jagged curves with no smooth components, such as the complicated Koch snowflake fractal curve named for Helge von Koch who explored the geometry in a 1904 paper.

locally a minimizing path; as a result, it is important in advanced mathematics and physics classes. Leonhard Euler published differential equations for geodesics in 1732. Mathematicians also investigated the classification of smooth curves. One invariant is the length of a curve. In general, length does not distinguish two different curves. It turns out that two other invariants, called the "curvature" and "torsion," work much better for this purpose. Broadly speaking, at any point on the curve, the curvature measures the deviance of the curve from being a straight line, and the torsion function measures the deviance of the curve from being a plane curve. Furthermore, the fundamental theorem of curves states that these invariants determine the curve, a result that is proved in twenty-first-century college differential geometry classrooms using the Frenet–Serret Formulas. These are named for Jean Frédéric Frenet and Joseph Serret, who independently discovered them in the nineteenth century.

With further investigations by prominent mathematicians, like Carl Friedrich Gauss, Gaspard Monge, Jean-Victor Poncelet, and their students, the theory of curves, particularly smooth curves, matured into an active field of research. The findings in the theory of curves not only enriched the realm of curve studies, they also contributed to the development of new ideas that ended up revolutionizing mathematics in the nineteenth century. Broadly speaking, the general definition of a curve is topological; namely, a curve is defined as a continuous map from an interval to a space. Curves can be algebraic (those defined via algebraic equations). For instance, a plane curve can also be expressed by an equation

$$F(x, y) = 0$$

and a space curve can be expressed by two equations

$$F(x, y, z) = 0 \text{ and } G(x, y, z) = 0.$$

A curve is algebraic when its defining equations are algebraic—a polynomial in x and y (and z). The cardioid, a heart-shaped curve whose Cartesian equation is

$$(x^2 + y^2 - 2ax)^2 = 4a(x^2 + y^2)$$

and the asteroid, whose Cartesian equation is

$$x^{\frac{2}{3}} + y^{\frac{2}{3}} = a^{\frac{2}{3}}$$

where a is a constant, are algebraic curves.

Before analytic geometry, each of these curves had been expressed using geometric investigations; for example, a circle turning around a circle that sweeps out the cardioid, or wheels turning within wheels that form the asteroid. Transcendental curves cannot be defined algebraically and include the brachistochrone curve, also known as the "curve of fastest descent"; very complicated looking fractal curves, such as the Koch snowflake, named for Helge von Koch, who explored the geometry in a 1904 paper; and paradoxical sounding space-filling curves, discovered by Giuseppe Peano in 1890. The last two types of curves can be extremely jagged curves with no smooth components.

An algebraic curve of the form $y^2 = x^3 + ax + b$, where a and b are real numbers, satisfying the relation $4a^3 + 27b^2 \neq 0$, is called an "elliptic curve." Geometrically, this condition ensures that the curve does not have any cusps, self-intersections, or isolated points. On the points of elliptic curves (including the point at infinity), one can define an operation by three points sum to zero, if and only if they are collinear. This interesting feature of elliptic curves, besides being an important algebraic structure to be studied on its own, also has found some astonishing applications, such as in cryptography for developing elliptic curve-based public-key cryptosystems. Elliptic curves are also important in number theory; they are effective tools in integer factorization problems. They also turned up as an instrumental tool in the proof of Fermat's Last Theorem, named for Pierre de Fermat.

Further Reading

Boyer, C. B. "Historical Stages in the Definition of Curves." *National Mathematics Magazine* 19, no. 6 (1945).

Lockwood, E. H. *A Book of Curves*. New York; Cambridge University Press, 1963.

National Curve Bank. http://curvebank.calstatela.edu/home/home.htm.

O'Connor, John, and Edmund Robertson. "MacTutor: Famous Curves Index." http://www-history.mcs.st-andrews.ac.uk/Curves/Curves.html.

Dogan Comez
Sarah J. Greenwald
Jill E. Thomley

Equations, Polar

Category: History and Development of Curricular Concepts.
Fields of Study: Algebra; Communication; Connections; Geometry; Representations.
Summary: Polar coordinate systems were developed in the seventeenth century and have numerous modern applications.

The polar coordinate system is a coordinate system for the plane in which each point is determined by a distance from a fixed point, called the "pole," and an angle from a fixed direction, called the "polar axis." In normal usage, the pole is analogous to the origin in the Cartesian coordinate system, named for René Descartes. Both polar and rectangular (Cartesian) coordinates require two bits of data to place a point in the plane. While the Cartesian coordinate system requires knowing and placing two chosen lines to serve as axes, polar coordinates requires knowing one fixed point and one fixed ray. This characteristic makes polar coordinates useful in navigation. Students in twenty-first-century high schools are introduced to polar coordinate systems and the topic is further developed in college mathematics and physics classrooms.

History

The concept of using an angle and a radius may be dated to the first millennium B.C.E. There are references to Hipparchus of Rhodes (c. second century B.C.E.) using a type of polar coordinates to establish the positions of the stars that he studied. Archimedes of Syracuse describes his namesake spiral in the book *On Spirals*, as where the distance from a given point depends on the angle from a given radius.

In a number of articles about the development of polar coordinates, most notably the 1952 article "Origin of Polar Coordinates" by Julian Lowell Coolidge, further development of polar coordinates was generated by studying the Archimedean spiral. According to Coolidge's history, the first mention should go to Bonaventura Cavalieri in his 1635 treatise *Geometria indivisilibus continuorum* in which he studies the spiral of Archimedes. Cavalieri studies the area inside the spiral and relates it to other known areas.

Like all good stories in the history of mathematics, this assertion is not without disagreement. In 1647, Grégoire de Saint-Vincent in his work *Opus Geometricum* claimed that he was familiar with the method and had sent his work to Christopher Grienberger in 1625. Grienberger had died in 1636, and the priority of the work was the subject of an article by Moritz Cantor in 1900.

Spiral curves were of interest to many mathematicians, including Gilles Personne de Roberval, James Gregory, Descartes, and Pierre Varignon. Gregory, Descartes, and Varignon all used a type of transformation of coordinates that heralded the complete development of polar coordinates. It appears to be Jacob Bernoulli and Isaac Newton who most completely developed these transformations. Bernoulli worked on the lemniscate and introduced the terms "pole" and "polar axis." Newton investigated transformations between coordinate systems, including polar coordinates, in his work *Method of Fluxions*, which was written in 1671 but not published until 1736.

Applications

Polar coordinates are the basis for navigation and radar, since the direction of travel can be given as an angle and distance from the origin. The radar screen that is used in air traffic control uses the location of the radar transmitter/receiver as the pole and magnetic north as the polar ray, zero degrees. This aspect and the fact that the angles continue in a clockwise direction instead of a counterclockwise direction are the major differences between a navigational use and the mathematical system. This same radar is the basis for all weather radar that is available for viewing either on television or from the Internet. Each radar location (there are 178 National Weather Service Doppler weather radar locations that cover the United States) sets a pole and covers a specific area. Storms are located and their paths are computed using the overlaps. This information must be transformed from the polar system (how far from the radar site and at what angle) into GIS coordinate system and then placed on a map to go to television or to the Internet. One well-known measuring device is the polar planimeter, created by mathematician and physicist Jacob Amsler in the nineteenth century. It measured the area enclosed by a curve. Amsler switched careers to focus on mathematical instruments, and he produced thousands of Amsler planimeters.

Other examples of the use of polar coordinates are very simplified uses in planning sprinkler systems in a building, as well as in irrigation systems in landscape and farming. Each of the sprinkler heads serves as a pole, and different walls, boundary lines and such serve as polar axes.

Different microphones have different recording patterns depending on the specific purpose. The omnidirectional microphone is used when sound from all directions is to be recorded, such as a choir or a large group. A cardioid microphone is a unidirectional microphone, which would be used to record a performer but not the crowd. Bidirectional microphones are used in an interview situation where the voices of both the interviewer and interviewee need to be recorded. The pattern of sounds that are picked up by the microphone are a lemniscate—the figure studied by Bernoulli.

Further Reading

Boyer, C. B. "Newton as an Originator of Polar Coordinates." *The American Mathematical Monthly* 56, no. 2 (1949).

Coolidge, J. L. "The Origin of Polar Coordinates." *The American Mathematical Monthly* 59, no. 2 (1952).

"A Periodic Shift in Polar Roses for Valentines Day." http://www.nikolasschiller.com/blog/index.php/archives/category/renderings/quilt/polar-coordinates.

DAVID C. ROYSTER

Exponentials and Logarithms

Category: History and Development of Curricular Concepts.
Fields of Study: Algebra; Communication; Connections.
Summary: Exponential and logarithmic functions are used to study and analyze a variety of mathematical relationships.

Much of the language and notation of mathematics involves a very advanced shorthand. As ideas grow and become more complex, mathematicians seek ways to express highly condensed thought in relatively simple terms. Exponents are an elementary example: if one wants to multiply the number 2 times itself 10 times, rather than write "2·2·2·2·2·2·2·2·2·2" one can write "2^{10}" instead. From these beginnings, which date to ancient Egypt and Babylon, the remarkable worlds of exponential and logarithmic functions emerge. When one develops the understanding of what it means to take 2 to any real number power, one naturally considers the function $f(x) = 2^x$, an example of what is called an "exponential function." For larger and larger positive x, the function grows amazingly fast: $2^{10} = 1024$, $2^{20} = 1,048,576$, and $2^{30} = 1,073,741,824$.

The exponential function $f(x) = e^x$, where e is the so-called "natural base," an irrational number whose decimal approximation is $e \approx 2.71828$, is an important exponential function. With e in homage to the great Swiss mathematician Leonhard Euler (1707–1783), this special exponential function $f(x) = e^x$ might rightly lay claim to the title of "the most important function in all of mathematics." Exponential growth and decay functions, along with the number e itself, have a wide variety of uses and applications.

In classrooms in the twenty-first century, the logarithm of a number is defined as the exponent or power to which a stated number, called the "base," is raised to obtain the given number. The development of logarithms in the seventeenth century led to a revolution in scientific calculation, especially when the slide rule replaced tables of logarithms. While the advent of calculators and computers eliminated the need for calculation by logarithms in the latter part of the twentieth century, logarithms remain important in order to understand financial and natural processes. For instance, the Richter scale to measure earthquakes, named for Charles Richter, is a logarithmic scale. In chemistry, the pH scale is based on the negative logarithm of the concentration of free hydrogen ions. Students in the middle grades investigate exponential notation while high school students explore exponential and logarithmic functions.

Archimedes of Syracuse investigated that the addition of what he called "orders" corresponded with their product, known today as the "first law of exponents." The number e may have first appeared in the early seventeenth century in an appendix to John Napier's work on logarithms. This number also arose in the work of Christiaan Huygens in the mid-seventeenth century when he was exploring the area under the hyperbola $xy=1$. Finally, in the late seventeenth century through work involving continuous compound interest, Jacob Bernoulli was led to consider the expression

$$\left(1+\frac{1}{n}\right)^n$$

for large values of n, and this expression approaches e as n grows without bound. Mathematicians explored many issues related to e and exponentials, including such people as Euler, Gotthold Eisenstein, and others, who investigated the convergence of sequences of iterated exponentials. Bernoulli may also have been the first mathematician to realize that the number e was intricately linked to emerging ideas with logarithms.

The Natural Exponential Function

Because any exponential function can be written in terms of e, one finds that functions of the form $P(x) = Me^{kx}$, where M and k are constants that depend on the context, arise in many natural settings. Exponential cell and population growth, as well as exponential decay in radioactive materials, are modeled by functions of this form. Once the values of M and k are identified, the function easily indicates the corresponding output for any input value x. For example, if a car is initially valued (at time $t = 100$) at $10,000 that depreciates at a certain continuous rate, one might use the function $P(t) = 10000e^{-0.2t}$ to model the worth of the car in year t.

Functions like this generate very natural questions, including ones like "At what time t will the car's value be $3,000?" Before trying to answer this more complicated question, consider some simpler ones. For instance, what value of t makes $10^t = 17$? Since $10^1 = 10$, while $10^2 = 100$, it seems like there ought to be a number between 1 and 2 such that 10 raised to that power is 17. But what is the number? Here, some very considerable mathematical ideas are involved: the function $y(t) = 10^t$ is continuous; the range of y is all positive real numbers; and $-y$ is always increasing, making it a one-to-one function. All these facts together combine to indicate that one can pick any positive real number y and know that there must be one and only one real number t that satisfies the equation $10^t = y$. In other words, there is a function h that takes any positive real number y, and to this value y associates the real number t so that 10 raised to the power t is y. This explanation is how teachers usually describe to students where logarithms come from—the logarithm is the very function that accomplishes this association. It is all a matter of perspective; if t is known and y is sought, the exponential function is used, while if y is known and t is sought, the logarithm function is used. Expressed in words, it is "y equals 10 to the power t" and "t is the power to which we raise 10 to get y." Babylonian clay tablets presented similar questions.

Historical Development

Historically, the further development of logarithms arose very differently. In the late fifteenth century and early sixteenth century, both John Napier and Jost Burgi, who were each interested in key problems in astronomy, developed logarithms for a much different use: as a new tool to help do arithmetic with large numbers. Their approach to logarithms was fundamentally geometric, as algebra was not yet sufficiently well developed to aid their work, although Napier's approach was more algebraic than Burgi's methods. Napier noted, "Seeing there is nothing that is so troublesome to mathematical practice, nor doth more molest and hinder calculators, than the multiplications, divisions, square and cubical extractions of great numbers, which besides the tedious expense of time are for the most part subject to many slippery errors, I began therefore to consider in my mind by what certain and ready art I might remove those hindrances." In 1624, Henry Briggs published logarithm tables in *Arithmetica Logarithmica* and he is noted by some as perhaps the man most responsible for popularizing logarithms among scientists. The development of the slide rule made logarithms easy to use, since they reduced the reliance on tables. In 1620, Edmund Gunter noted logarithms on a ruler by marking the position of numbers relative to their logarithms. William Oughtred placed two sliding logarithmic rulers next to each other and by 1630, the portable circular slide rule reduced multiplication computations to the act of lining up two numbers and reading a scale. Logarithms remain a useful way to deal with large numbers in the early twenty-first century, because the logarithm of a large number is a much, much smaller one. R. C. Pierce Jr. noted, "It has been postulated that logarithms literally lengthened the life spans of astronomers who had formerly been sorely bent and often broken early by the masses of calculations their art required." Modern mathematicians have also come to fully understand the connection between logarithms and the area under the curve $xy = 1$, which was explored by Huygens in the 1600s.

Using Logarithms to Solve Exponential Functions

Perhaps the most powerful property of logarithms is that they "undo" exponential functions. For example, for the natural logarithm of base e, denoted "ln," one obtains $\ln(e^5) = 5$. Remember, $\ln(e^5)$ means "the power to which one raises e to get e^5." This power, of course, is 5. The general property that holds here is that for any real number t, $\ln(e^t) = t$. This rule proves to be immensely useful in solving exponential equations. To see how, consider an earlier example: the function $P(t) = 10000\, e^{-0.2t}$ (the value of a car in year t). At what time t will the car's value be $3,000? This question is equivalent to solving the equation:

$$0.3 = e^{-0.2t}.$$

Taking the natural logarithm of both sides of the equation "undoes" the effects of the exponential function and hence gains more direct access to the variable t: $\ln(0.3) = \ln(e^{-0.2t}) = -0.2t$.

Dividing both sides of the last equation above by -0.2, one finds that

$$t = \frac{\ln(0.3)}{-0.2} \approx 6.0199$$

so that the car's value will be $3,000 in just over six years. The natural logarithm of 0.3 is central to answering the question.

While the motivation for the need for logarithms can be seen in relatively elementary terms—solving exponential equations—the actual mathematics that explains what logarithms really are and how they work is deep and is best supported using some sophisticated ideas from calculus. Even with exponential functions, there are some big questions without answers: how is e to the 5th power calculated? How is the natural logarithm of 0.3 computed? Until the invention of personal computers in the 1970s, such computations were all done by hand, usually with the assistance of elaborate tables, or with slide rules. At one point in history, entire books were written that held nothing but tables of values for logarithms. People now use inexpensive hand-held calculators, computer algebra systems like *Maple* or *Mathematica*, or even Google, and each returns a value almost immediately. These modern technological tools rely on a rich and beautiful mathematical theory of exponential and logarithmic functions. Beyond their interesting mathematical properties, exponential and logarithmic functions remain important for their many applications, such as the key role that exponential functions play in the study of differential equations, including those that model vibrations in bridges and buildings, thus forming a central component of modern civil engineering.

Further Reading

Maor, Eli. *e: The Story of a Number*. Reprint. Princeton, NJ: Princeton University Press, 1994.

Nahin, Paul J. *An Imaginary Tale: The Story of i (The Square Root of Minus One)*. Princeton, NJ: Princeton University Press, 2010

Pierce Jr., R. C. "A Brief History of Logarithms." *The Two-Year College Mathematics Journal* 8, no. 1 (1977).

Stoll, Cliff. "When Slide Rules Ruled." *Scientific American* 294, no. 5 (2006).

Strogatz Steven. "Power Tools—NYTimes.com" http://opinionator.blogs.nytimes.com/2010/03/28/power-tools.

Matt Boelkins

Function Rate of Change

Category: History and Development of Curricular Concepts.
Fields of Study: Algebra; Calculus; Communication; Connections.
Summary: The rate of change of a function is a key focus of differential calculus.

Though calculus has a reputation for impenetrability compared to algebra and geometry, one of its two main branches, differential calculus, is concerned with derivatives, or rates of change. Rate of change is intuitively understood: the stock market is falling, but how fast? The rolling ball is slowing down, but when will it stop? A derivative is the rate of change of a mathematical function, discovered through a process called differentiation.

History and Language of the Study of the Rate of Change

The ancient Greeks wrestled with the concept of change. Parmenides of Elea asserted that change is impossible, while Heraclitus of Ephesus believed that everything changes and nothing remains still. Aristotle accepted some forms of change but he denied a "change of change" related to motion. Historians have commented that a lack of recognition of a rate of change of a velocity was the major stumbling block to the development of calculus by Archimedes of Syracuse almost 2000 years before Sir Isaac Newton.

The language that describes changing quantities can still be confusing. A politician announces that a proposed spending bill will "cut the deficit" because it will "lower the rate at which the deficit is growing." Geologists announce that global oil production continues to increase and that "the rate at which production is growing is decreasing." Does this mean that the rate of global oil production itself may soon also decrease? Reflecting on a particular twenty-first century recession, economists shared that "the rate at which jobs are being lost is decreasing." Are these announcements good news or bad news?

Average and Instantaneous Rates of Change

In order to understand these statements, several key ideas about functions must be investigated; in particular, it is important to know what it means to say that a function is "increasing" or "decreasing," as well as whether a function's rate of change is increasing or decreasing. A familiar physical situation is helpful to consider. Let a ball be tossed into the air and follow its height above the ground at time t.

In the late 1600s, Sir Isaac Newton correctly conjectured that the ball's height can be modeled by a certain quadratic function. Consider the function $y = h(t) = -16t^2 + 32t + 4$, where h is measured in feet and t is measured in seconds, and observe its graph below. The goal is to study how the function changes as time moves forward and, hence, to understand what is meant formally by "the rate of change of the function."

From our understanding of quadratic functions, we can observe that at time $t = 0$, when the ball is tossed, its height is $h(0) = 4$ feet. The ball lands when its height is $h = 0$, which occurs for the positive value of t that satisfies $-16t^2 + 32t + 4 = 0$; the quadratic formula indicates that the positive t that satisfies this equation is $t = 1 + \sqrt{5}/2 \approx 2.118$ seconds.

Finally, since the vertex of the parabola occurs at $t = -32/(2 \cdot (-16)) = 1$, it follows that the maximum height the ball reaches is $h(1) = -16 \cdot 1^2 + 32 \cdot 1 + 4 = 20$ feet. Clearly the ball is going up on the interval from $t = 0$ until $t = 1$, and the ball is going down thereafter. Perhaps a more interesting question is "*how* is the ball going up and going down?" Or, "how fast is the ball rising or falling at a particular moment?" For instance, consider the interval $[\frac{1}{2}, 1]$.

On that interval, the ball rose 4 feet, since $h(1) - h(1/2) = 20 - 16 = 4$. In addition, half a second of time elapsed. This knowledge shows that the function's "average rate of change" on the time interval $[\frac{1}{2}, 1]$ is

$$\frac{h(1) - h(\frac{1}{2})}{1 - \frac{1}{2}} = \frac{20 - 16}{\frac{1}{2}} = 4 \times \frac{2}{1} = 8 \text{ feet per second.}$$

The units on this quantity are important: the numerator is measured in feet, while the denominator is in seconds, so the overall units are "feet per second," reflecting the rate of change of height with respect to time. The algebraic form of the average velocity on the time interval $[a, b]$,

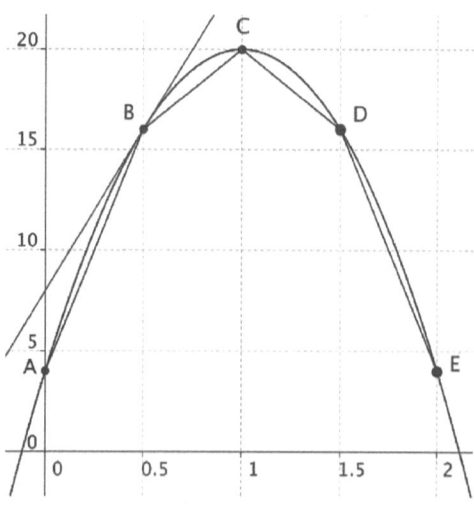

$$\frac{h(b)-h(a)}{b-a}$$

is reminiscent of another familiar quantity: the slope of a line that passes through the points (x_1, y_1) and (x_2, y_2) is given by

$$\frac{y_2 - y_1}{x_2 - x_1}$$

Straight line segments are used to model and approximate the parabolic function, for example, from point B to point C in the graph. Hence, it is seen that the average rate of change of the function h on a given interval is understood visually to be the slope of a line that passes through two points on the graph of h.

The average rate of change of the function quantifies how fast the ball is rising or falling and will vary on different intervals. This accurately reflects that the ball is "falling faster," since its average rate of change is more negative than on the preceding interval. For instance, on the interval $[0, \frac{1}{2}]$, the average rate of change of h is

$$\frac{h(\frac{1}{2}) - h(0)}{\frac{1}{2} - 0} = \frac{16 - 4}{\frac{1}{2}} = 12 \cdot \frac{2}{1} = 24 \text{ feet per second,}$$

while on the interval $[\frac{1}{2}, 1]$, the average rate of change is

$$\frac{h(1) - h(\frac{1}{2})}{1 - \frac{1}{2}} = \frac{20 - 16}{\frac{1}{2}} = 4 \cdot \frac{2}{1} = 8 \text{ feet per second.}$$

These quantities confirm numerically what we can see visually from the graph: the ball is rising faster during the first half-second than it is during the second half-second. What happens in the subsequent half-second?

Similar computations, shows that on the time interval $[1, \frac{3}{2}]$, the average rate of change is

$$\frac{h(\frac{3}{2}) - h(1)}{\frac{3}{2} - 1} = \frac{16 - 20}{\frac{1}{2}} = -4 \cdot \frac{2}{1} = -8 \text{ feet per second.}$$

Here, for the first time, a negative average rate of change is encountered; the minus sign is extremely important, as it is the numerical indicator that the ball is falling. From the symmetry of the parabola, one can expect (and can calculate to check) that on $[\frac{3}{2}, 2]$, the average rate of change of h is −24 feet per second. This result accurately reflects that the ball is "falling faster," since its average rate of change is more negative than on the preceding interval.

It is next natural to seek to understand the difference between the ball's average rate of change on a time interval and its "instantaneous" rate of change at a single value of t. By taking average rates of change on smaller and smaller time intervals, one encounters a remarkable phenomenon. For instance, consider the average rates of change on $[0.5, 1]$, $[0.5, 0.6]$, $[0.5, 0.51]$, and $[0.5, 0.501]$. The average rate is 8 feet per second on the first interval; on the next interval, the function's average rate of change is

$$\frac{h(0.6) - h(0.5)}{0.6 - 0.5} = \frac{17.44 - 16}{0.1} = 14.4 \text{ feet per second.}$$

On $[0.5, 0.51]$, similar computations reveal that the average rate of change is 15.84 feet per second, while on the final interval, $[0.5, 0.501]$, the rate is 15.984. Here, despite the fact that one is dividing by numbers that are getting closer and closer to zero (0.5, 0.1, 0.01, 0.001), it can be seen that the resulting quantities themselves seem to be settling down, nearer and nearer a single number. Calculus is the mathematics that allows these ideas to be made precise. The notion of limits and other key related ideas allow mathematicians to move from the notion of average rate of change to instantaneous rate of change and indeed the instantaneous rate of change of the ball's height with respect to time at the time $t = 0.5$ is 16 feet per second. By considering the corresponding line segments that pass through two points on the curve, the so-called secant lines actually approach a single line that is "tangent" to the curve at the point (0.5, 16), as pictured in the graph as point B. The line touches the curve only at (0.5, 16), has slope 16, and represents the instantaneous rate of change of the ball's height with respect to time at the moment $t = 0.5$.

The Beginnings of Calculus

This idea of moving from average rates of change to instantaneous ones is the starting point for the entire subject of differential calculus. Abu Arrayhan Muhammad ibn Ahmad al-Biruni investigated instantaneous velocity and acceleration approximately 1000 years ago, and Isaac Barrow may have been the first to draw tangents to curves in 1670. The development of calculus led to a rich collection of concepts centered on the idea of a rate of change, many of which were introduced by Isaac Newton in his attempts to develop a universal theory of gravitation. For instance, Newton attempted to avoid the use of the infinitesimal by forming calculations based on ratios of changes and he determined the area under a curve by extrapolating the rate of change. In fact, Newton's second law states that the rate of change of momentum of a body is equal to the force acting on the body in the same direction. Gottfried Leibniz also investigated concepts related to rate of change. He explored maxima, minima, and tangents in 1684, but mathematicians had difficulty understanding his six-page work.

Other notions of rate of change are also important in mathematics and in real-life applications. For example, in 1847, Jean Frédéric Frenet assigned a frame of vectors to each point on a curve and described the twists and turns of the curve by the rate of change of the frame. Joseph Alfred Serret independently considered similar ideas in 1851. The Frenet–Serret frame continues to be useful in the early twenty-first century when it is impossible to assign a natural coordinate system.

Many real-life problems, such as population growth, can be expressed and modeled as an equation involving a quantity and its rate of change. All of these ideas rest in some way on the fundamental concept of slope, which is investigated beginning in the middle grades, while the notion of rate of change is first developed in high school. Other methods to solve these types of problems are studied in the field of differential equations, which is usually introduced in college.

Applications of Rates of Change

Returning to two of the original questions about the meaning of certain statements and whether they are good news or bad news: a proposed spending bill will "cut the deficit" because it will "lower the rate at which the deficit is growing." This is not great news, since the deficit is still growing, but a deficit growing at a decreasing rate is better than one growing at an increasing rate. It would be much better to hear that the budget deficit itself was decreasing. Next, the information that "the rate at which oil production is growing is decreasing" may likely mean that the rate of oil production could be leveling off and soon start to decrease—the concept of "peak oil" (when the rate of daily global oil production reaches its maximum)—and many analysts believe humans have just passed this peak and that the rate of oil production will only continue to fall from here.

With the Earth's human population growing at a present rate of 83 million people per year, as well as so many other changing quantities, collective efforts to understand resource allocation and management require sound understanding of rates of change and trends in data. Calculus and its language of change are a key tool in building a sustainable future for humanity and the planet.

Further Reading

Cohen, David, and James Henle. *Calculus: The Language of Change*, Sudbury, MA: Jones and Bartlett, 2005.

Connally, Eric, et al. *Functions Modeling Change: A Preparation for Calculus*. 2nd ed. Hoboken, NJ: Wiley 2007.

Dunham, William. *The Calculus Gallery*. Princeton, NJ: Princeton University Press, 2005.

Strogatz, Steven. "Change We Can Believe In." *New York Times* (April 11, 2010). http://opinionator.blogs.nytimes.com/2010/04/11/change-we-can-believe-in.

Matt Boelkins

Functions

Category: History and Development of Curricular Concepts.
Fields of Study: Algebra; Communication; Connections.
Summary: There are many different types of functions that arose in a variety of historical contexts.

A mathematical function expresses the idea that one quantity can be completely determined by another quantity. As such, a function is a well-defined rule between two sets, A and B, where each element x of A is assigned to one element y of B. The careful study of the implications and applications of this definition comprises much of mathematics. Functions are ubiquitous throughout nearly all fields of mathematics and their importance cannot be overstated. Since a function expresses a relationship between an independent variable and a dependent variable, many real-world phenomena are modeled using functions. Functional correspondence between variables can be expressed verbally, algebraically, or graphically. Although the formal definition of a function is a relatively recent development, this concept has been implicit since the beginning of mathematics.

Examples of Functions

Functions can be classified in various ways and many of these specific classes of functions are the bases of entire fields of study within mathematics. Below are a few examples to illustrate some areas encompassed by the definition of a function. The algebraic equation $y = x^2$ expresses y as a function of x. No matter what value has been chosen for x, only one y value will be produced.

The expression $y = \pm x$ does not represent y as a function of x. If $x = 1$ is entered into the equation, then two y values are obtained, $y = -1$ and $y = 1$.

The expression $y = \cos(x)$ is a trigonometric function. For any given angle expressed in radian measure, the cosine function uses a rule from trigonometry to produce a real number from -1 to 1.

An example of an exponential function is given by the formula $y = 2^x$. For instance, when $x = 2$, $y = 4$. Exponential functions are useful for situations in which the rate of growth of a population is directly proportional to the size of the population, such as modeling the growth of bacteria or the decay of a radioactive sample.

A piecewise function is defined by different formulas for different values of x that are entered into the function. The Heaviside function is an example of a piecewise function: $y = 1$ for $x \geq 0$, and $y = 0$ for $x < 0$.

If one assigns "on" to 1 and "off" to 0, this function models the behavior of turning on a switch at $x = 0$ and leaving it on. It is also used in the study of electric circuits to indicate the surge of an electric current.

A barcode scanner at a supermarket can be thought of as a function. After a particular barcode is scanned, only one price will be displayed.

A computer program that obtains the five-digit zip code for an address acts as a function because every address in the United States has only one zip code assigned to it.

Equivalent Formulations

A helpful way to think about a function is as a machine that produces exactly one output y for each input x. This "machine" could simply be a description or a list of the pairings between x and y. Although this may be easier conceptually, in practice it is unfeasible to list all of the pairings when there is a large number (or even an infinite number) of x values. When one considers an infinite number or a large number of x values, it is more advantageous to have a mathematical formula that precisely relates x and y.

Graphs of Functions

Another way to represent a function is by using a graph. One begins with a set of x values from any subset of the real number line and a function. The function then specifies a y value for each x. This results in a collection of pairings (x, y). Each of these pairs denotes a point, which is plotted on the xy-plane in the two-dimensional Cartesian coordinate system. The collection of all points produced by this process is the graph of the function.

By virtue of the definition of a function, every x value on the graph is paired with, at most, one y value. Thus, any vertical line that is drawn will cross the graph of the function at most one time. This is known as the "vertical line test": a curve in the xy-plane is the graph of a function if—and only if—no vertical line crosses the curve more than once. As a consequence of the vertical line test, given a curve, it is relatively easy to determine if it is the graph of a function. All nonvertical straight lines are graphs of functions. Circles are not graphs of functions.

History

The notion of a function has been implicit throughout the history of mathematics. Addition is the most fundamental arithmetical operation and, although it was not initially formulated as such, it is a function of two variables. The pair of numbers to be added is the input and the resulting sum is the output of the addition function. Ancient cultures, such as the Babylonians, developed extensive tables of mathematical calculations of the reciprocals and square roots of positive whole numbers. These calculations involve specific functions but were not formulated using the function concept.

In the fourteenth century, Nicole Oresme had a rudimentary grasp of the idea that one changing quantity can be dependent upon another. He depicted this relationship graphically using a method he called the "latitude of forms." This depiction was the first known attempt of the graphical representation of a function. Throughout the Middle Ages, the latitude of forms continued to be studied; however, further development of the function concept was hampered by the absence of a suitable algebraic framework.

The formal study of functions began in the late seventeenth century with the discovery of calculus. Although in 1692, Gottfried Leibniz first introduced the word "function" in association with the tangent problem in calculus, its first definition did not emerge until 1718 with Johann Bernoulli. The primary representation of a function at that time was from curves that were connected to physical problems.

As the eighteenth century unfolded, algebraic equations were increasingly being used to represent two-dimensional curves. As a result, the emphasis and focus of the function concept evolved from a graphical setting to that of algebra. This shift was evident in Leonhard Euler's 1748 treatise on functions, *Introductio in analysin infinitorium*. Euler's definition of a function was that of an "analytic expression" or an algebraic formula that could contain any combination, the five arithmetic operations: exponentials, logarithms, trigonometric ratios, derivatives, and integrals. Euler's emphasis on the algebraic formulation of functions was evident, as the first volume of the *Introductio* contains no graphs.

The middle of the eighteenth century saw a development of the function concept when a controversy arose over the solution to the Vibrating String Problem. Given an elastic string with fixed endpoints and deformed into an initial shape, the string was released and began to vibrate. The problem was to determine a function that would describe the shape of the string at any future time. In 1747, Jean Le Rond d'Alembert produced a solution in the form of an algebraic equation. A year later, Euler verified that this solution was correct but he disagreed that it was the most general. He claimed that d'Alembert had neglected several initial shapes of the string that could be drawn freehand and for which there were no algebraic expressions. Euler also pointed out that other initial shapes could be obtained by piecing together simpler curves. This critique led to the acceptance of functions produced from freehand drawing for which there may not be any algebraic formula and piecewise defined functions.

Another solution to the Vibrating String Problem further complicated matters. In 1753, Daniel Bernoulli solved the problem differently than Euler and d'Alembert and arrived at a seeming contradiction: different mathematical expressions defined the same function. The controversy was not resolved at the time; it remained for Joseph Fourier to expand upon Bernoulli's idea. Fourier's solution to the Heat Conduction Problem in 1807 resulted in a revolution of the understanding of a function. Fourier demonstrated that a function could be expressed as an infinite series of sine and cosine functions, now known as a Fourier series. These series demonstrated that two different expressions could define the same function. Following the development of Fourier series, the connection between the geometric and algebraic forms of a function was further solidified. Furthermore, ideas from calculus were reexamined in a new light.

In 1837, Lejeune Dirichlet suggested a definition of "function" that was closely related to the modern definition. Dirichlet emphasized that a function provides a relationship between two variables but allowed for freedom in describing the rule that describes how x and y are related. To show how pathological a function can become, Dirichlet introduced the function $y = c$ for x an irrational number, and $y = d \neq c$ for x a rational number.

This badly behaved function cannot be sketched and there is no algebraic equation defining it.

Since the nineteenth century, it was a natural evolution to recast Dirichlet's definition by using set theory. The modern definition for a function now provides a correspondence between two sets, which may or may not be numerical; for example, functions between algebraic structures like groups or geometric objects like surfaces.

Further Reading
Boyer, Carl B. *A History of Mathematics*. Princeton, NJ: Princeton University Press, 1985.
Kleiner, Israel. "The Evolution of the Function Concept: A Brief Survey." *College Mathematical Journal* 20, no. 4 (1989).
Stewart, James. *Calculus: Early Transcendental*. 6th ed. Belmont, CA: Thomson Brooks/Cole, 2008.

Courtney K. Taylor

Graphs

Category: History and Development of Curricular Concepts.
Fields of Study: Communication; Connections; Geometry.
Summary: Graphs and diagrams are one way to represent mathematical information and may convey it more clearly than other methods or reveal interesting patterns and relationships.

Graphical representations have been found since antiquity in such places as cave drawings and maps. Modern graphs are fundamental to the organization and presentation of information. The concept of a graph developed along with advances in printing, mathematical theory, and empirical observations, especially in such fields as astronomy, cartography, chemistry, crystallography, calculus, geometry, probability, and statistics. Quantitative information, such as data points or functions, is often exhibited and analyzed in graphs. In twenty-first-century classrooms, students of all ages explore various types of graphs. Graph theory is a branch of mathematics that studies mathematical graphs in which vertices or nodes, representing objects, are connected by edges that represent relationships between the objects.

Early Graphs

Attempts to depict familial relationships led to a variety of family trees and graphs, some of which survived from the Middle Ages. Family trees have long been of historical and personal interest in tracing ancestry and nobility relationships. The rise of genealogical social networks at the beginning of the twenty-first century led to huge family trees. Researchers and software developers have created new ways to visually represent ever-changing family relationships, including divorce and remarriage.

Some graphical representations arose in the context of puzzles or games. For instance, variants of a game known as Men's Morris have long appeared in carvings on Roman buildings and in cathedrals in medieval England. In the thirteenth century publication of Alfonso X of Castile, the *Libro de los Juegos* (*Book of Games*), an illustration shows a Morris game board with nodes that represent the positions of game counters and connections between them that represent the moves. The beginnings of graph theory are often attributed to eighteenth-century mathematician Leonhard Euler. In 1736, he presented a solution showing that it was impossible to continuously traverse the seven bridges of Konigsberg, Russia, without retracing the same path or lifting the writing utensil. However, his paper does not contain any graphs, although it does contain maps of Konigsberg. Continuous figure tracing also appeared in Danish folk puzzles, as well as in the Angola, Zaire, and Zambia region in Africa, and in the New Ireland and Vanuatu regions in Oceana. Euler also did not use graphs in his 1759 work on a Knight's Tour, where a knight must traverse each square on a chessboard without repetition. In 1771, Alexandre-Theophile Vandermonde used a graph drawing in this context.

One common notion of a graph is a pictorial representation of a function. The graph of a function passes the vertical line test, so that each input has one assigned output. Egyptologists Somers Clarke and Reginald Engelbach noted that an ancient Egyptian architect's diagram showed a curve with vertical lines and coordinate measurements expressed in units of cubits, palms, and digits. The graphical depiction of changing quantities where one quantity depends on another can be found in the fourteenth-century publications of Nicole d'Oresme and in *De latitudinibus formarum* (the Latitudes of Forms), which may also have been written by d'Oresme. The development of coordinate geometry, coordinate axis systems, and the notion of a function in the seventeenth and later centuries, through the work of René Descartes, Pierre de Fermat, Gottfried Leibniz, Peter Dirichlet, and others, allowed for the graphical representations of algebraic formulas, curves, and other mathematical objects. Thomas Hankins noted that graphs started appearing in 1770 in the context of

> . . . the statistical atlases of William Playfair, the indicator diagrams of James Watt and the writings of Johann Heinrich Lambert. . . . That leaves us with the question of what is to count as a graph. If we include maps and geometrical and astronomical diagrams, graphs are very old indeed. What was new in the late eighteenth century was a diagram with rectangular coordinates that showed the relationship between two measured quantities. Lambert called them *Figuren*, Watt called them "diagrams," and William Playfair called them "lineal arithmetic." William Whewell, who seemed to

rename everything that he came into contact with, called them the "method of curves."

Gaspard Monge's eighteenth-century work also influenced the development of graphs as well as fields like architecture and engineering. He is known as the "father of descriptive geometry," which studies three-dimensional geometry through two-dimensional images.

The earliest known uses of the terms "graph," "graph paper," and "graph theory" originated in the nineteenth and twentieth centuries. Mathematician James Sylvester is noted as the first to use the term "graph" in the publication *Nature* in 1878 when he described a chemical graph. Graphs in chemistry originated earlier, such as in the eighteenth century when chemist William Cullen referred to an "affinity diagram" to model molecular forces. Alexander Brown depicted molecules as graphs in 1864. Mathematician Arthur Cayley developed graph theory in the 1870s in the context of chemistry. Some have cited Julius Peterson's late-nineteenth-century work as the start of the field of graph theory. The Peterson graph that is named for him is explored in graph theory classes. George Chrystal referred to the "graph of a function" in his 1886 algebra text: "This curve we may call the graph of the function." Graph paper was originally known as "squared paper" or "coordinate paper" and was patented by Dr. Buxton in the late eighteenth century. The use of "graph" as a verb may date to an 1898 work on applied mechanics, in which John Perry advised: "Students will do well to graph on squared paper some curves like the following ... in each case calculate y. Plot the values of x and y as co-ordinates of points on squared paper, and draw the curve passing through the points...."

Types of Graphs

The study of logical statements, their implications, and their relationships resulted in a variety of different types of diagrams. Young children use Venn diagrams, which represent set containments and intersections using overlapping circles. These were named for philosopher and mathematician John Venn because of his nineteenth century work to formalize and generalize them. The concept of a Eulerian Circle, named for Euler, is related. Aristotle's square of opposition is named for the ancient Greek philosopher. Aristotle analyzed deductive logic among various statements. Fourth-century mathematician and philosopher Anicius Manlius Severinus Boethius also explored the logical relations.

In some versions, the square of opposition was presented as a square diagram that contained propositions that were represented inside circles. Lines that connected the circles represented the relationships between the propositions. College students and researchers in fields like logic, topology, algebra, and geometry use commutative diagrams with arrows or other symbols to represent mappings or logical relationships.

Educational Graphs

Students in the twenty-first century investigate a wide variety of graphical and diagram representations. In primary schools in the United States, students represent and analyze problems using graphs, charts, data, and functions; the graphs also serve as a subject of study themselves. William Playfair's 1786 publication *The Commercial and Political Atlas* is noted as the beginning of charts, such as bar charts and line charts, and perhaps the first appearance of statistical time series graphs. He also invented the pie chart in 1801. In the middle grades, students also generalize patterns with graphs and identify and contrast linear and nonlinear graphs.

In addition, they convert between symbolic algebraic formulas and graphical representations and learn about graphical features, such as the slope or intercept of a line and the changing quantities in a graph. In high school, students continue to create graphical representations and they approximate the rate of change of a function from its graph. In calculus, students use graphs to further understand the properties of functions, such as their derivatives, integrals, and the notion of concavity. The integral is defined as the area under a curve, and students use Riemann sums, named for nineteenth-century mathematician Bernhard Riemann, to approximate the area using rectangles.

The widespread use of graphing calculators and computer software in the late twentieth century changed the way that students explored graphs. They were able to quickly graph complex equations and large amounts of data to look for patterns. Students and teachers explore candidates for categories like the most beautiful graph, the funniest graph, or the worst graph, which some define as the most misleading and others as the most confusing.

Debate continues regarding what is the desired balance between by-hand graphing skills versus a reliance on graphical methods on the computer or calculator. Some teachers argue that if students do not understand how to create graphs, they will not be able to fully understand misrepresentations or analyses. Another area that has taken on new prominence in twenty-first century schools and colleges is discrete mathematics and graph theory.

Further Reading

Ascher, Marcia. "Graphs in Cultures: A Study in Ethnomathematics." *Historia Mathematica* 15 (1988).

Biggs, N., E. Keith Lloyd, and R. Wilson. *Graph Theory 1736–1936*. New York: Oxford University Press, 1999.

Friendly, Michael. "Milestones in the History of Thematic Cartography, Statistical Graphics, and Data Visualization." http://www.math.yorku.ca/SCS/Gallery/milestone/milestone.pdf.

Hankins, Thomas. "History of Science Society Distinguished Lecture: Blood, Dirt, and Nomograms, A Particular History of Graphs." *Isis* 90 (1999).

Kruja, Eriola, Joe Marks, Ann Blair, and Richard Waters. "A Short Note on the History of Graph Drawing." 9th International Symposium on Graph Drawing. Berlin: Springer-Verlag, 2002.

Sarah J. Greenwald
Jill E. Thomley

Infinity

Category: History and Development of Curricular Concepts.
Fields of Study: Calculus; Communication; Connections; Number and Operations.
Summary: Infinity is an important part of the curriculum and has a rich and interesting history.

Counting comes naturally to humans. Children as young as 2 years old begin to associate numbers with groups of objects: 1, 2, 3, 4, and 5 are quickly understood. The concept of "plus one" also develops early in life. Given any whole number, there is always a "next number," the one achieved by adding one. As such, early in life we face the reality that there is no largest number, for given a number of any size, adding one to it produces a number that is yet bigger. That is, the set of natural numbers {1, 2, 3, 4, 5, . . .} is infinite. While the concept of infinity is fundamental in mathematics, cosmology, and theology, many of the advances in understanding infinity were met with severe criticism or worse. For example, according to some stories, Hippasus, a member of the Pythagorean order, was drowned for divulging the existence of infinite non-repeating decimals. Revolutions in philosophy and mathematics resolved many of the fascinating paradoxes related to infinity, but infinity continues to challenge and interest us today.

A Hotel Example

Infinity is a concept, but it is not itself a number. To illustrate how the notion of infinity is different, it is helpful to turn to one of the great mathematicians of all time, David Hilbert (1862–1943). In 1900, he spoke to the International Congress of Mathematicians about 23 unsolved problems that he considered to be the most important to the progress of mathematics—the search for solutions to these problems shaped a great deal of twentieth-century mathematics, and some even remain open to this day. Besides being a leading mathematician, Hilbert was also a thoughtful teacher, and he was reputed to have used the following paraphrased story to challenge his students to think about the curious nature of infinity.

> A mathematician owned an unusual hotel, one with infinitely many rooms. Each room was assigned a natural number—Room 1, Room 2, and so on—and on one occasion, it happened that every room in the hotel was filled. A customer seeking a room walked into the lobby and asked the manager if there were any openings. The manager reported that every room was full but that there was a way for the customer to get a room.
>
> The occupant of Room 1 was asked to move to Room 2; the occupant of Room 2 moved to Room 3; and in general, the person in Room N stepped next door to Room $N + 1$. The customer who had requested a room at the entirely full hotel was now able to occupy Room 1.

The next day, when the hotel was still completely full, an unusual charter bus arrived, carrying infinitely many passengers, all seeking rooms. At first, the members of this group were disheartened to learn that the hotel was completely booked. But the mathematically savvy manager once again had a solution.

The occupant of Room 1 was asked to move to Room 2; the occupant of Room 2 moved to Room 4; the person in Room 3 went to Room 6; and in general, the person in Room N stepped down the hall to Room $2N$. The customers getting off the bus were now able to move into all of the odd-numbered rooms, as rooms 1, 3, 5, 7... were all open.

While this story may seem far-fetched because there are only a finite number people alive on Earth, it illustrates some remarkable properties of natural numbers and raises concerns, such as whether more natural numbers exist than there are natural numbers.

Infinite Sets

Georg Cantor's revolutionary ideas on the sizes of such infinite sets form the basis of many ideas in modern mathematics, including the fields of analysis and calculus. For example, removing the odd natural numbers from the set of all natural numbers

$$\{1, 2, 3, 4, 5, 6 \ldots 2n, 2n+1, \ldots\}$$

leaves the set $\{2, 4, 6 \ldots 2n, 2n+2, \ldots\}$ which is yet another infinite set. Galileo Galilei believed that the sizes of infinite sets could not be compared or contrasted. However, Cantor and mathematicians today agree that since a first even natural number can be identified, a second even natural number, and so on, just as a first natural number can be identified, a second natural number, and a third, then there are the same number of even natural numbers as there are natural numbers since they can be put in one-to-one correspondence. Cantor also proved that there are uncountable sets that have a different measure of infinity, such as the real numbers. However, Cantor did not receive the recognition during his lifetime that he has today. Some theologians believed his work challenged the uniqueness and infinity of God, and both mathematicians and theologians strongly objected to his work at the time.

Limits

A question that has intrigued many people over the centuries is whether or not the numbers 1 and $0.\overline{9}$ are the same. In fact they are, as the following argument shows. If we consider the number $0.\overline{9}$, observe the following:

$$0.\overline{9} = 0.9999\ldots = \frac{9}{10} + \frac{9}{100} + \frac{9}{1000} + \frac{9}{10000} + \cdots$$

Certainly, two numbers can be added, three numbers, four numbers, and indeed, as many finite numbers as likened be. From this, observe the following:

$$\frac{9}{10} + \frac{9}{100} = \frac{99}{100};$$

$$\frac{9}{10} + \frac{9}{100} + \frac{9}{1000} = \frac{999}{1000};$$

$$\frac{9}{10} + \frac{9}{100} + \frac{9}{1000} + \frac{9}{10000} = \frac{9999}{10000}; \text{ and,}$$

$$\frac{9}{10} + \frac{9}{100} + \frac{9}{1000} + \cdots + \frac{9}{10^n} = \frac{10^n - 1}{10^n}.$$

Since this last sequence of numbers converges to the number 1, one concludes that the infinite sum is 1. That is,

$$0.\overline{9} = 0.9999\ldots = \frac{9}{10} + \frac{9}{100} + \frac{9}{1000} + \frac{9}{10000} + \cdots = 1.$$

At first glance, this may seem strange to a person unaccustomed to the role of limits in mathematics. But, as was perhaps first understood by Archimedes in antiquity, limits are the bridge from the finite to the infinite, and they are indispensable to mathematics and the mathematician. Understanding infinity allows for the understanding that the numbers 1 and $0.\overline{9}$ are the same.

Paradoxes

Certainly the concept of infinity presents some challenges and unusual situations. Greek philosopher Zeno of Elea was known for posing paradoxes that challenged mathematicians for centuries. For instance, in Zeno's Paradox, a person walks toward a wall by each time stepping half the remaining distance, thus taking time stepping half the remaining distance, thus taking an infinite number of steps but (theoretically) never actually reaching the wall.

Another example is Gabriel's Horn, an infinite surface that can be easily generated by revolving a simple curve about an axis. Interestingly, the surface is not named after its discoverer, Italian physicist and mathematician Evangelista Torricelli, but is rather named after the Archangel Gabriel in order to connect the infinite with theology. This infinite surface can be shown to contain finite volume yet have infinite surface area. In other words, Gabriel's Horn, if filled with paint, would require only a finite volume, yet that paint could not cover the surface of the horn. While situations like these initially seem impossible, mathematics provides interesting and satisfying explanations of these phenomena.

Modern Developments

In the twentieth and twenty-first centuries, mathematicians continue to grapple with the concept of infinity. French mathematicians Émile Borel, René Baire, and Henri Lebesgue explored rationalist ideas, while a group of Russian mathematicians led by Dmitry Egorov linked mathematics to philosophy and theology. Building upon the French work and using mystical insights gained during their religious practice of Name Worshipping, they founded descriptive set theory, which transformed mathematical analysis.

However, this did not resolve the contradictions of infinitesimals in calculus, which Sir Isaac Newton, Gottfried Leibniz, and Bishop Berkeley had wrestled with during the development of that subject in the seventeenth and eighteenth centuries. Abraham Robinson created the field of nonstandard analysis in 1960 when he gave a rigorous definition of an infinitesimal number, and mathematicians continue to explore the implications of both standard and non-standard analysis. Besides there being infinitely many natural numbers, there are even infinitely many prime numbers. Primes form the building blocks of numbers and in many ways the very foundation of mathematics. In a similar way, calculus rests upon the notion of limit, which at its core involves infinite processes. Because so much of the subject naturally involves the infinite, mathematicians have had to face, understand, and conquer infinity; more than this, the presence of infinity in the world guarantees that there will always be more mathematics to explore, discover, and comprehend.

Further Reading

Clegg, Brian. *A Brief History of Infinity: The Quest to Think the Unthinkable*. London: Robinson, 2003.

Graham, Lauren. *Naming Infinity: A True Story of Religious Mysticism and Mathematical Creativity*. Cambridge, MA: Belknap Press of Harvard University Press, 2009.

Rucker, Rudy. *Infinity and the Mind*. Princeton, NJ: Princeton University Press, 2004.

Stillwell, John. *Roads to Infinity: The Mathematics of Truth and Proof*. Natick, MA: A K Peters, 2010.

MATT BOELKINS

Limits and Continuity

Category: History and Development of Curricular Concepts.
Fields of Study: Algebra; Calculus; Communication; Connections.
Summary: One of the key concepts of calculus, the limit is the value a function approaches as its input approaches a given value.

The concepts of limit and continuity are fundamental in calculus, analysis, and topology. Their inception can be traced back more than 2000 years to Greece, China, Babylon, Egypt, and other places. During the inception of calculus, introduced independently by Isaac Newton and Gottfried Wilhelm Leibniz, these concepts were still vague and controversial. In the twenty-first century, these concepts are explored in high school. The modern limit of a function $f(x)$ is the value the function tends to when x changes in a structured way; for instance, x approaches a specific value. Many different definitions of limits in calculus can be combined into a single definition of limit in topology. Moreover, a function is continuous if it preserves closeness or, equivalently, if it preserves limits, and topology is the study of continuous functions and the properties they preserve. These ideas underpin many mathematical results and can be used to organize and simplify mathematical processes. Limits are also useful in real life to understand such concepts as demographics, finance, or terminal velocity. Modeling discrete data with a continuous function and the notion of continuous payments are also important. Leibniz defined a principle of continuity, or *lex continuitatis*, which inspired philosophers such as Charles Peirce. The notions of limit and continuity are still debated philosophically, as in whether growth spurts are continuous over time.

The Ideas of Limit and Continuity in the Ancient World

The idea of limit in the ancient world was related mainly to two activities: one was more practical, like measuring length, area, and volume, and the other was more abstract, such as making sense of numbers that are not rational. For example Archimedes from Greece and Liu Hui in China used regular polygons, inscribed in a circle, increasing the number of sides of the polygons, in order to compute the length of the circumference and the area of a circle. In the process, approximations of π were computed. Eudoxus from Greece created his theory of proportions to legitimize irrationals like $\sqrt{2}$.

This theory is expounded in Book Five of Euclid's *Elements*. It is also a precursor of the contemporary theory of the real numbers. Ancient mathematicians, such as Zeno of Elea and Aristotle, wrestled with the notion of continuity. They debated whether motion, time, and space are continuous. The paradox about Achilles and the tortoise illustrated the interplay between the ideas. The paradox states that Achilles can never overtake the tortoise if the tortoise is given a head start, because by the time Achilles reaches its initial position the tortoise has farther advanced and so on; infinitely many segments of time are necessary.

The Calculus of the Infinitesimals

In the middle of the seventeenth century after significant advances in science, particularly in physics, mechanics, and geometry, the methods of infinitesimal calculus were introduced independently by Isaac Newton in England and by Gottfried Wilhelm Leibniz in Germany. Newton assumed that geometric magnitudes are generated by continuous motion, and some historians suggest that he may have been the first to present a limit argument using an infinitesimal like *epsilon*. Leibniz explored a principle of continuity but is not thought to have explored the derivative as a limit. He viewed the ocean as continuous.

The quest to find an acceptable, rigorous foundation for the new calculus was ongoing. There were attempts made to follow the method of exhaustion from the Greeks or to use series instead of infinitesimals to introduce the derivative. Jean le Rond d'Alembert stressed the importance of a firm foundation for limits and explored a geometric limit of secant lines. But it was Augustin-Louis Cauchy who introduced contemporary definitions of limit and continuity and placed them as a cornerstone of calculus. It is interesting to mention that another mathematician, Bernard Bolzano, from Prague and a contemporary of Cauchy, came up with the same definitions first, but he was more isolated and his work did not get the same recognition that the *Cours d'analyse* enjoyed. The German mathematician Karl Weierstrass solidified the rigorous definitions and made significant contributions to the development of analysis.

Contemporary Definitions

The limit of a function is a dynamic concept. The input of the function varies and the output varies as well. Intuitively speaking, one says that $\lim_{x \to c} f(x) = L$ when the values of f become closer and closer to the number L when c gets closer and closer to c. This intuitive concept is easy to grasp and also not difficult to observe if one has a graph of the function f. But this should be expressed rigorously, so that there is a tool to verify whether the limit exists.

Definition: The $\lim_{x \to c} f(x) = L$ if and only if for each $\varepsilon > 0$ there is a $\delta > 0$, such that if $0 < |x - c| < \delta$, then $|f(x) - L| < \varepsilon$.

This definition enables mathematicians to verify the existence of limit or to make an argument that there is no limit and is also a tool to prove many properties about limits. The number c does not have to be in the domain of the function, but one should be able get δ close to it from the domain for any positive δ. The concept of limit is used to define a continuous function.

Definition: A function f is continuous at a point c if c is in the domain of the function and for any positive ε there is a positive δ, such that if $|x - c| < \delta$, then $|f(x) - f(c)| < \varepsilon$.

The concept of limit is used to define the definite integral and to measure area.

Contemporary Developments.
Limits of other objects, such as sequences and geometric spaces, can be defined and are important in many disciplines of mathematics and continuity is still explored in the field of topology. One twentieth-century development that goes back to the history of limits occurred around 1960 when Abraham Robinson entertained the idea that the advantages of infinitesimal calculus can be utilized as soon as the infinitesimals are defined in a rigorous way. This would eliminate the use of limit in the way it is known and make the analysis very much like algebra, as soon as the number system is extended to permit infinitely small and infinitely large numbers. This is exactly what Robinson did using tools from logic, a development called "non-standard analysis."

Further Reading
Edwards, Charles. *The Historical Development of Calculus*. New York: Springer, 1994.
Ilarregui, Begoña, and J. Nubiola. "The Continuity of Continuity: A Theme in Leibniz, Peirce, and Quine." In *Leibniz und Europa, VI.* Hannover, Germany: Gottfried-Wilhelm-Leibniz-Gesellschaft, 1994.

ELENA TONEVA

Linear Concepts

Category: History and Development of Curricular Concepts.
Fields of Study: Algebra; Communication; Connections.
Summary: Linear relationships are a fundamental concept of mathematics.

From ancient civilizations to modern societies, people use linear concepts in a multitude of ways, including statistical analysis, for advanced mathematics, and in scientific applications, many of which are designed to solve real-world problems. The fundamental idea of a linear relationship involves comparing two or more quantities which form a straight line when graphed. A basic linear relationship comparing two quantities is represented by the linear equation $y = mx + b$, where x and y are the quantities which vary in direct proportion to each other, m represents the slope of the line, and b represents the value where the line crosses the y-axis on the coordinate plane. This idea implies that the quantities in a linear relationship depend upon each other. One of the most common ways to represent a linear relationship is the linear equation. Linear equations are equations involving one or more unknown values, called "variables," which are of the first degree. Linear equations are the simplest type of equation, since the unknown quantities are always raised to the power of one. Spaces of lines, such as a plane, are also the simplest type of geometric space. However, linear equations and the methods of finding the solutions become more complex as the number of unknowns increases or when solving more than one linear equation at a time. Students begin to explore linear concepts in the primary grades, and these are built upon and extended throughout high school and college.

Linear equations are used extensively in applied mathematics, particularly in modeling and representing real-world phenomena. Linear relationships are also used in advanced mathematical applications and modeling, typically by reducing nonlinear equations to linear equations or by constraining events within a set, or "system," of linear equations. The development of general methods for solving linear equations was a slow process because of the limitations of communicating and representing the unknown quantities in linear relationships. These equations were initially solved using the elementary operations of addition, subtraction, multiplication, and division. The problems and their solutions were written using words. Algebraic methods of solving equations were not developed until a system of symbolic notation replaced the use of words. The modern practice of using variables in place of unknown values did not gain widespread use until the sixteenth century. Before that time, problems were typically written using only words or by using a limited set of symbols.

Linear Equations
The earliest known linear equations and methods of solving them are found in several ancient civilizations. These societies used linear equations and systems of linear equations to solve problems arising in everyday life, particularly based on civic and government needs. Although the historical information and records that exist from these ancient civilizations are fragmented, there exists enough evidence to show how the Babylonian, Egyptian, Chinese, and Islamic

civilizations used and solved linear problems. Within the Babylonian civilization, the need for computational techniques beyond simple counting arose in areas of commerce, taxation, and construction. The Babylonians wrote their problems on clay tablets and included many examples of solving linear equations and systems of linear equations. These numerical problems were expressed rhetorically, without symbolic notation, and were provided to show the method of solution for a particular example. Reasons and explanations were not given, nor were any general methods of solution. The Egyptians were also concerned with commerce, taxation, and construction. They also described the methods used to solve linear equations arising from everyday life, such as dividing loaves of bread or a given amount of grain. They wrote their problems on papyrus, which was made from a reed plant, very few of which exist today. The most famous surviving papyri are the Rhind Papyrus and the Moscow Papyrus. One particular procedure the Egyptians devised for solving linear equations is known as the "method of false position." This procedure began with guessing a value for the unknown and then adjusting the value until the correct result was found. This method was also used by other ancient civilizations and continued to appear in elementary algebra textbooks until the nineteenth century. The Chinese used a similar method of false position but made two guesses for the unknown rather than one guess. This method is known as the "method of double false position" and was later used by Islamic mathematicians. This approach continued to be used in Europe until the 1600s when advances in symbolic notation made solving linear equations a much more simple process. It took many years for a symbolic system to develop and allow for the development of general solutions to linear equations.

Linear Modeling

Linear relationships appear extensively in modeling applications. In statistics, simple linear regression is commonly used to model the relationship between two variables when that relationship appears to be generally linear. That is, when plotted on the coordinate plane, the data tend to cluster about a straight line. It can also be used to make predictions for situations when the value of one variable is known and the other is not. The concept of linear regression was developed by Sir Francis Galton in the late nineteenth century while investigating genetic inheritance. A description of this method was published in the article "Regression Towards Mediocrity in Hereditary Stature." The term "regression" was actually a reference by Galton to observed effects in the data, not to the method itself, yet the statistical process of fitting a line to data still bears this name. An extension of the method, called "multiple linear regression," is used to model relationships between several variables in n dimensions.

Many types of advanced mathematical models also rely on the use of linear relationships. For example, linear programming is used in business applications as a way to model important decisions that lead to maximum profit. This modeling is accomplished by constraining the variables, such as production costs, within a system of linear equations. Linear programming is also used in a modeling process known as "linear optimization." This modeling process has a wide variety of applications in areas such as business, finance, engineering, and industry. Linear programming and linear optimization are based on the mathematical procedure of defining all of the related variables as linear relationships. This process was first developed in the 1940s and is one of the few mathematical applications that has a wide range of practical uses as well as a theoretical development of the mathematics.

Many definitions in mathematics rely on linear approximations. The derivative of a function of one variable at a point is the slope of the tangent line (the slope of the line that best approximates the curve at a given point). Mathematicians such as Isaac Barrow and Sir Isaac Newton made linear concepts a fundamental part of their work in the development of calculus. In higher dimensions the derivative is a linear transformation that is represented as a matrix. In geometry, a surface is defined as a space that locally looks like a plane. Georg Friedrich Bernhard Riemann defined higher dimensional spaces, now called "manifolds," as locally looking flat and possessing shortest paths that are straight. In 1917, Albert Einstein used Riemann's mathematics in order to present a model for the universe that was consistent with his theory of relativity.

Linear Algebra

Linear algebra is a subject that is fundamental to modern mathematics and applications. It arose from the study of coefficients of systems of linear equations, and linear concepts are fundamental in this area. For example, Arthur Cayley explored linear maps or transformations, and Giuseppe Peano was the first to give an abstract definition of the algebraic structure of linear vector spaces.

The late development of a symbolic notational system used in solving linear equations slowed the development of finding general methods for solving these equations. The first breakthrough in using algebraic techniques to solve linear equations occurred in the sixteenth century when Jacques Peletier proposed a general rule of algebra. This general rule involved setting up linear relationships as equations and finding the roots, a method still used in the teaching of algebra. With the adoption of a system of symbolic notation, the applications of linear equations continue to evolve and to be used in numerous ways, from basic equation solving to advanced mathematical techniques in both pure and applied mathematics. Linear algebra has a long history in mathematics, and linear concepts are considered one of the most important concepts in mathematics because of their appearances in so many levels of both pure and applied mathematics and in a multitude of real-world applications.

Further Reading

Bressoud, David. *The Queen of the Sciences: A History of Mathematics.* Chantilly, VA: Teaching Company, 2008.

Coxford, Arthur, ed. *The Ideas of Algebra, K–12: 1988 Yearbook.* Reston, VA: National Council of Teachers of Mathematics, 1988.

Katz, Victor. *A History of Mathematics.* Boston, MA: Addison-Wesley, 2009.

Kelli M. Slaten

Mathematical Certainty

Category: History and Development of Curricular Concepts.

Fields of Study: Problem Solving; Reasoning and Proof; Representations.

Summary: Mathematics is arguably the most stable and rigorous source of knowledge; yet any system of mathematical reasoning is incomplete.

At one end of the spectrum in mathematics, it seems as if people can be absolutely certain, using mathematical proofs of concepts. At the other, in statistics and many applied mathematics fields, it is virtually impossible to be certain, but people can make probabilistic statements with regard to degrees of uncertainty inherent in a given calculation or statement.

The Greeks may have been the first to attempt a rational explanation of nature. The crucial tool in their investigations was mathematical reasoning. They assumed that all questions about nature can be answered by reason and that all these answers are knowable and can be discovered, a property known as "completeness." They also assumed that all answers are compatible, which is called "consistency." However, the evolution of mathematical certainty revealed that these assumptions can never be fully realized. The notion of what is certain and what is uncertain is a fundamental component that is threaded in various ways throughout twenty-first-century mathematics curricula. For example, in primary school, students investigate the differences between "likely" and "unlikely" events. Students also develop inductive and deductive reasoning by exploratory investigations and examples as well as by proofs.

Axiomatic Systems

The Pythagoreans (c. 585–500 B.C.E.), a school influenced by Pythagoras of Samos, offered a mathematical plan of nature. The Greeks' goal was to rationally explain why things are the way they are. They confronted a fundamental question: can all knowledge be verified? Aristotle (384–322 B.C.E.) answered "no," since there are self-evident truths (called "axioms") that cannot be explained. Moreover, in geometry, Aristotle said a proposition is proven when it is shown to logically follow from the axioms and other proven propositions. Euclid of Alexandria (323–285 B.C.E.) knew of these developments and incorporated them into his text, *Elements*. It is recognized as the prototype for how mathematics should be done: well-thought out axioms, precise definitions, carefully stated theorems, and logically coherent proofs.

Formulation of the axioms or "postulates" (the Greek term for axioms about geometry) is the crucial step in building an axiomatic system. These statements should be intuitively self-evident, and, from these, it must be possible to deduce the important properties of the objects of study. Later, mathematicians found assumptions used in *Elements* that were not explicitly stated in the axioms. Credit for completely and successfully axiomatizing Euclid's geometry is generally given to David Hilbert.

The Parallel Postulate and Non-Euclidean Geometries

Euclid's fifth (or parallel) postulate states that, "through a given point, not on a given line, only one parallel line can be drawn to the given line." Almost immediately, this postulate was controversial. Many did not find it to be self-evident and thought it required a proof. Over two millennia, countless mathematicians tried to derive the parallel postulate from the others, all with no success. These futile efforts had important consequences in all of mathematics. Beginning in the eighteenth century, some mathematicians began to use indirect methods to "prove" this postulate.

Though unsuccessful, the indirect methods led to the discovery of non-Euclidean geometries—using the other axioms but denying the parallel postulate. Attempts to prove non-Euclidean geometries were invalid were essentially attempts to show that they were inconsistent. Eventually, it was determined that Euclidean and these other geometries were consistent and complete. The discovery of non-Euclidean geometries revealed that mathematics could deal with completely abstract axiomatic systems, which no longer had to correspond to beliefs based on real-world experiences.

The Consistency and Completeness of Mathematics

It became clear that the most important considerations for an axiomatic system were its consistency and completeness, and at the International Congress of Mathematicians in 1900, Hilbert addressed these problems. He felt all mathematics should be put on a sound basis using the axiomatic method. In 1904, Hilbert constructed an arithmetic model of Euclidean geometry, showing that geometry was a subset of arithmetic. Mathematicians then set out to show the consistency of arithmetic, from which it would follow that Euclidean geometry was consistent.

These efforts ended in 1931 with the results of Kurt Gödel. His first Incompleteness Theorem showed that in any axiomatic system rich enough to include the arithmetic of the natural numbers, it is possible to prove some statements that are false, showing the system is inconsistent; or it is not possible to prove some statements that are true, showing the system is incomplete. In his second Incompleteness Theorem, Gödel showed the question of whether an axiomatic system is consistent cannot be determined within the system. Gödel's results revealed that any mathematical reasoning system based on axioms as rich as arithmetic can never be fully realized—such systems must be either incomplete or inconsistent. Modern mathematicians operate under the assumption that mathematics is incomplete and not inconsistent.

Different Axioms Lead to Different Mathematics

One axiom of set theory, the axiom of choice (AC), was used implicitly for years before it was explicitly described. The AC states that for any collection of nonempty mutually exclusive sets, finite or infinite, there is a set that contains exactly one element from each set. The AC with Zermelo–Fraenkel (ZF) axiomatic set theory, named for Ernst Zermelo and Abraham Fraenkel, is the basis of modern mathematics. In 1938, Gödel proved that if ZF set theory without the AC is consistent, then ZF set theory with the AC is also consistent. So, just as it is possible to choose between different acceptable geometries in which the parallel postulate may or may not be true, it is possible to choose between different acceptable ZF set theories in which the AC may or may not be true. On one hand, theorems requiring the AC are fundamental in such areas as modern analysis. Then again, by adopting the AC, results such as the Banach–Tarski paradox, named for Stefan Banach and Alfred Tarski, can be derived, which says a golf ball can be divided into a finite number of pieces and then rearranged to make a solid sphere the size of the Earth. Thus the decision as to which axiomatic system to adopt cannot be made lightly—different mathematics can be derived from these different axiomatic systems.

Valid Proofs

The idea of a valid proof depends on one's philosophical approach to mathematics. A number of schools of thought have evolved, including (1) the logistic school, which holds mathematical proofs derive from logic; (2) the intuitionist school, which maintains mathematics takes place in the human mind and is independent of the real world—it composes truths rather than derives implications of logic; (3) the formalist school in which proofs follow from the application of a system of axioms; and (4) the set theoretical school, which derives proofs from the axioms of set theory.

Mathematics remains our most rigorous form of knowledge. As proofs grow more complicated, mathematicians worry they will have to accept a greater degree of uncertainty in solutions. For example, the entire proof of the Classification Theorem for Finite Simple Groups consists of an aggregate of hundreds of research papers and over 10,000 printed pages. Additionally, the Four-Color Problem solution has been achieved only on the computer and involves checking a prohibitively large number of cases. Some mathematicians believe that since it is not reasonable and possible for any one individual to check all these cases, then a valid proof has not been provided for such problems.

Further Reading

Borba, M. C., and O. Skovsmose. "The Ideology of Certainty in Mathematics Education." *For the Learning of Mathematics* 17, no. 3 (1997).

DeLong, Howard. *A Profile of Mathematical Logic*. Reading, MA: Addison-Wesley, 1970.

Kline, Morris. *Mathematics: The Loss of Certainty*. New York: Oxford University Press, 1980.

Wainer, Howard. *Picturing the Uncertain World: How to Understand, Communicate, and Control Uncertainty Through Graphical Display*. Princeton, NJ: Princeton University Press, 2009.

Linda Becerra
Ron Barnes

Mathematical Modeling

Category: History and Development of Curricular Concepts.
Fields of Study: Communication; Connections; Problem Solving; Representations.
Summary: Modeling reformulates scenarios to mathematical elements for analysis and problem solving.

Mathematical modeling has been in use since prehistory and was likely the first kind of mathematics ever employed. Mathematical modeling can be thought of as the activity involved in finding a solution to a real-life problem by working with a mathematical structure that captures the important characteristics of the situation. In the twenty-first century, mathematical modeling is found in many areas of mathematics, engineering, science, social science, and business and has often resulted in the formation of recognized "subdisciplines" within these fields. Research and applications occur in a diverse range of theoretical and real-world problems, and modeling is used in schools starting in the primary grades to help students visualize and solve problems, create alternative representations of various concepts, and make connections between different areas of mathematics. The advent of computers has facilitated mathematical modeling and allowed researchers to conduct simulations or find numerical solutions to problems that may be difficult to solve analytically. However, there are those who argue against overuse of mathematical models, citing issues of faulty data, unwarranted extrapolation, and the inherent error of attempting to quantify many complex or qualitative real-world phenomena. These types of criticisms have been applied to models associated with the financial and housing crises of the early twenty-first century and evidence on both sides of the global warming debate.

Anyone who has ever attempted to solve a story problem has dabbled in modeling. Consider the following story problem:

> I asked my dad for some money. He gave me 24 coins with three times as many dimes as quarters, for a total of $3.30. How many of each coin do I have?

To solve the problem, one converts the verbal statements into equations. The set of equations is the mathematical model, which can be solved to determine an answer. When formulating the mathematical equations, assumptions would be made, such as which denominations of coins to include in the model.

Process

Modern treatises on the modeling process often portray the steps involved in modeling using a diagram similar to Figure 1.

Starting with a problem statement, the first action is to determine the assumptions that should be made, information in the problem that is extraneous and can be neglected, quantities that are known (parameters) and unknown (variables), and relationships between the quantities. This work may entail using a variety of strategies, including developing or using existing physical laws, proportionality arguments, equations from the current experts in the field, or equations empirically determined from experimental data.

That first step will lead to a mathematical representation of the real-world situation. The mathematics may take the form of equations, inequalities, recursive relations, matrices, graphs, integrals, differential equations, geometric structures, or other mathematical objects.

The next action is to "solve" the mathematics, leading to an answer. That answer may be an exact solution, a simulation, or an approximation. The answer must then be interpreted in the context of the problem, and any approximations must be checked (validated) to see if the solution is correct. Lastly, the explanation of the answer in the context of the situation should be used to verify or predict the solution to the problem.

In practice for complex real-world problems, the modeling process is really a cycle that is traversed repeatedly as the model is refined to produce more realistic behavior. It is common to find that these processes involve multidisciplinary teams of professionals, including mathematicians and scientists, who participate in a dialogue to clarify and refine the assumptions based on the success of the last step in the process: analyzing the mathematics in the context of the problem.

History

There is archeological evidence from more than 10,000 years ago of simple mathematical ideas being developed to solve problems related to counting objects, measuring land area and distance, and recording time. More complex mathematical problem solving appears around 3000 B.C.E., when the cultures of Asia, the Middle East, and North Africa began using arithmetic, algebra, and geometry to solve problems in astronomy, building construction, and financial situations, such as taxation. The design and construction of complex pyramids and temples and the development of sophisticated astronomical calendars in Central and South America in the first century C.E. point to the development of mathematical ideas to solve problems. It can be argued, in fact, that most of the mathematics developed before 1800 was conceived to help model a situation in the real world.

Before the mid-nineteenth century, many of the real-world problems that were approached using mathematics would be classified today as astronomy, physics, or engineering. For example, Archimedes (287–212 B.C.E.) was instrumental in modeling physical tools, such as levers, and in the development of models for hydrostatics (the properties of water at rest, such as pressure). Eratosthenes (276–194 B.C.E.) used a geometric model and his knowledge of how the sun casts shadows to determine the circumference of the Earth. Abu Ali Hasan Al-Haitham, known more commonly as Alhazan (965–1040), developed the first principles of optics for spherical and parabolic lenses. Blaise Pascal (1623–1662) developed the ideas fundamental to probability while helping a gambling friend by modeling a dice-rolling game. Isaac Newton (1642–1727) is perhaps the best-known "mathematical modeler" who ever lived, famous for his ground-breaking work on the classical laws of motion and gravitation. Building on equations of fluid flow developed by Leonhard Euler (1707–1783), Claude Henry Navier (1785–1836), and George Stokes (1819–1903) produced the Navier–Stokes equations, which model velocity, pressure, temperature, and density of a moving fluid. The Navier-Stokes equations, a set of nonlinear partial differential equations, were truly understood only after the advent of modern digital computers in the 1960s.

Figure 1. The modeling process.

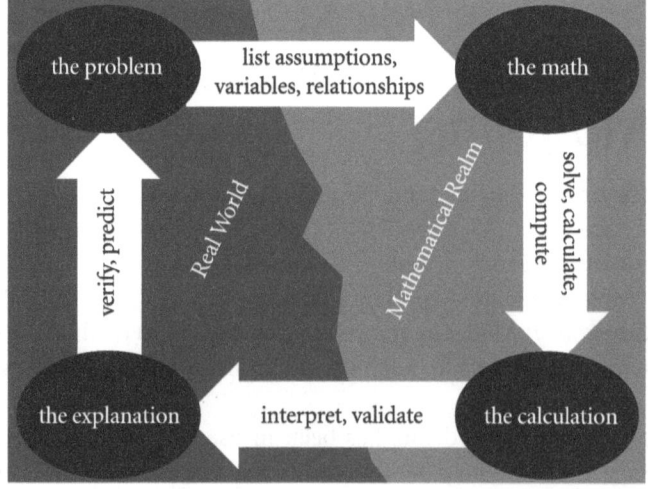

Figure 2. Malthus's exponential population growth model (left) as compared to the Verhulst model, which incorporates intra-species competition for resources.

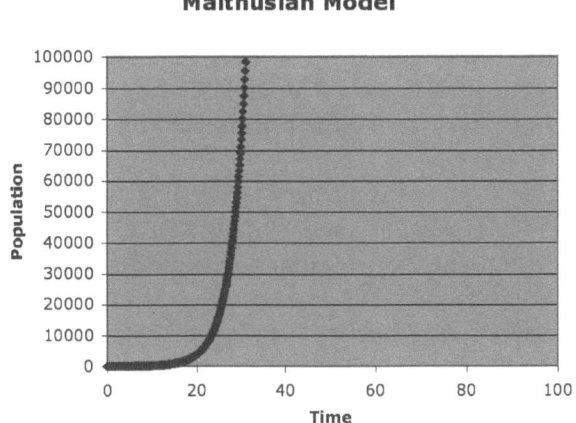
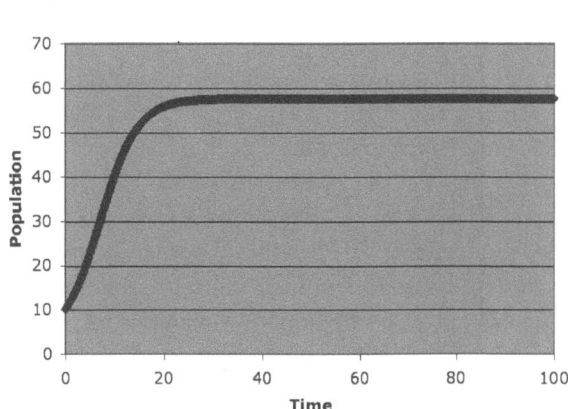

The nineteenth century saw an expansion into biological and social science modeling. Thomas Malthus (1766–1834) wrote about population growth and the familiar exponential model for population growth is named after him. Pierre Verhulst (1804–1859) took Malthus's ideas and developed the logistic, limited growth model (see Figure 2).

Late Nineteenth Century Through Twentieth Century

From the late nineteenth century forward, mathematicians have become more concerned with the development of theoretical—sometimes called "pure"—mathematics: abstract structures derived from fundamental axioms and built through proving theorems following logical precepts. However, this interest in mathematics for its own sake did not slow down the development and use of mathematics as a tool to model the real world. The application areas have become increasingly diverse, and the twentieth century saw the process of mathematical modeling adopted in many fields outside physics and engineering.

In the first two decades of the twentieth century, Albert Einstein (1879–1955) developed his theories of relativity, mathematical models that predict gravitational processes on the planetary scale more accurately than Newton's—now called "classical"—mechanics. Alfred Lotka (1880–1949) and Vito Volterra (1860–1940) worked in the 1920s on models of the interaction between predator and prey species, each arriving at the same model using different assumptions and arguments about how variables interact. Population models continue to be explored and refined through the present day. George Danzig (1914–2005) developed the simplex algorithm in 1947 to solve the mixing, supply chain, and other logistical problems that arose in World War II; these problems could be modeled with the well-understood linear programming approach, but the problems had so many variables and constraints that they were too complex to solve without computers. Linear programming is arguably the mathematical model most used in business and agriculture today. Edward Lorenz (1917–2008) developed one of the first nonlinear models for the atmosphere in the early 1960s, a precursor to the sophisticated climate models of today. His model displayed a very interesting sensitivity to initial conditions, and the study of this and similar models led to the field of chaos theory.

Twenty-First Century

In the twenty-first century, the use of mathematical modeling is ubiquitous across many research areas and academic disciplines. The Society for Mathematical Psychology publishes research in mathematical models used to examine psychological problems in neurology and cognition. The Society for Mathematical Biology concerns itself with applications of mathematics to modeling complex ecological systems, genetics, medicine, and cell biology. *The Journal of Mathematical Chemistry* is published by Springer-Verlag to provide a venue for researchers to share results from mathematical models of molecular behavior and chemical reactions. The American Sociological Association Section for Mathematical Sociology meets regularly to share research that uses "the language of mathematics to describe the structure, explain the events, and predict the dynamics of the social world." The American Institute of Physics publishes the *Journal of Mathematical Physics*, which focuses on applications of mathematical modeling to classical mechanics and quantum physics. The field of operations research, also called "management science," has evolved with the goal of solving mathematical models to determine the best business decision (often the maximum profit or minimum cost) given a situation in which there are limited resources. The Institute for Operations Research and the Management Sciences (INFORMS) is one of many organizations that publish results from this discipline.

The research presented through these venues tackles a diverse range of real-world problems. In medicine, mathematical models for physical principles of flow and pressure have been adapted and expanded to model the heart as a double-chambered pump. The flow of blood through the vessels can be examined and the parameters for flexibility of the vessels can be changed to investigate health conditions, such as hardening of the arteries brought on by aging. The ideal gas law and equations governing transport have been used to model how the lungs function to transport oxygen from the air inhaled to the blood in the aveoli in exchange for carbon dioxide. In the social sciences, Markov processes (mathematical matrices of transition probabilities) have been used to model social mobility and vacancy chains (the notion that a vacancy in a company or a house causes a chain reaction as others move in to fill the vacancy) as well as recidivism (the likelihood that a criminal will become a repeat offender). Generalizations of the Navier–Stokes and Lorenz models have been used to model the Earth's atmosphere, including sources of pollution and other greenhouse gases in an effort to prove and to predict the presence (or absence) of global warming. Scientists at NASA's Goddard Institute for Space Studies conduct research in three-dimensional atmospheric circulation models and in coupled atmosphere-ocean models in an effort to understand climate sensitivity.

With the advent and development of computers, increasingly sophisticated situations can be modeled and approximate solutions or simulations obtained using numerical algorithms. With modern computers to do the heavy computational work solving or simulating the mathematics, the most challenging step in the process is often the formulation of the mathematical model.

Further Reading

Beltrami, E. *Mathematical Models in the Social and Biological Sciences*. Boston: Jones and Bartlett Publishers, 1993.

Friedman, A., and W. Littman. *Industrial Mathematics: A Course in Solving Real-World Problems*. Philadelphia: Society for Industrial and Applied Mathematics, 1994.

Hadlock, C. *Mathematical Modeling in the Environment*. Reston, VA: Mathematical Association of America, 1998.

Hoppensteadt, F., and C. Peskin. *Modeling and Simulation in Medicine and the Life Sciences*. 2nd ed. New York: Springer-Verlag, 2002.

Klamkin, M., ed. *Problems in Applied Mathematics: Selections from SIAM Review*. Philadelphia: Society for Industrial and Applied Mathematics, 1990.

NASA Goddard Institute for Space Sciences. "GISS Research: Global Climate Modeling" (2010). http://www.giss.nasa.gov/research/modeling.

Smith, S. *Agnesi to Zeno: Over 100 Vignettes From the History of Math*. Berkeley, CA: Key Curriculum Press, 1996.

Swetz, F., ed. *From Five Fingers to Infinity: A Journey Through the History of Mathematics*. Chicago: Open Court Publishing, 1994.

HOLLY HIRST

Mathematician Defined

Category: Mathematics Culture and Identity.
Fields of Study: Communication; Connections.
Summary: Mathematicians work in a variety of fields and contribute widely to society.

Broadly construed, a mathematician is anyone who actively researches or studies mathematics. Many mathematicians work in academia as professors, involved in teaching, new research, or (most commonly) a combination of both. However, mathematicians are also employed in large numbers by industry, and there are innumerable amateur mathematicians who are drawn to mathematics, pursuing its study and research as an avocation. Some mathematicians, called "applied mathematicians," use mathematical ideas to solve problems arising in other disciplines; others, "pure" or "theoretical mathematicians" focus on furthering mathematics for its own sake. Of course, many mathematicians belong to both categories. The image of the mathematician is somewhat stereotyped in popular culture, but, in fact, mathematicians comprise an extremely diverse group. Mathematicians are women and men, girls and boys, old and young, and come from every country and culture.

If mathematicians are introduced to other mathematicians, they are unlikely to describe themselves as just "mathematicians." More often they would use a more precise term indicating their primary research interests, such as "number theorist," "analyst," "algebraist," "combinatorialist," "probabilists," or "logicians." The degree of specialization varies widely from one mathematician to the next. A mathematician may have only one area of research interest or may work across several. It is now seen as impossible for any single person to be expert in all areas of mathematics, but there are still so-called "generalists" who work in as many branches of mathematics as possible.

Mathematicians, Scientists, and Poets

While there are some overlap and blurred boundaries between the terms "mathematics" and "science" as these terms are used in ordinary discourse, the terms "mathematician" and "scientist" are usually used with more clearly distinct meanings. Scientists apply the scientific method, a continual process of investigating phenomena, collecting empirical data, formulating explanation hypotheses, and testing them by experiment; for scientists, experiments and empirical data provide the ultimate test of a theory. While mathematicians may also use experiments as part of their work, this is chiefly as a source of inspiration, as an aid in formulating conjectures and understanding complex concepts. Some might make the distinction that the scientist is generally an inductive reasoner, while the mathematician is generally a deductive reasoner. However, many applied mathematicians and statisticians may be more like scientists in this regard. To many people, it might seem that mathematicians and poets are polar opposites, or at the very least unrelated. It is remarkable, as such, how often great mathematicians and great poets speak of the vocations as intertwined. For example, Russian mathematician Sonia Kovalevsky (1850–1891) wrote, "It is impossible to be a mathematician without being a poet in soul." Likewise German mathematician Karl Weierstrass (1815–1897) wrote, "A

Rényi's and Erdös's Joke

A well-known joke definition among the mathematics community says that "A mathematician is a device for turning coffee into theorems." This aphorism evokes the image of mathematicians holed up alone in offices, drinking coffee by the pot as they fill blackboards and notebooks with computations and it is often attributed to the famous mathematician Paul Erdös. Though Erdös certainly did popularize this quip, it is more likely that it originated with his friend and colleague Alfréd Rényi. This notion comes from the then-nascent popularity of coffeehouses as gathering places for the European mathematicians. The stimulative effects of caffeine on the mathematician's brain, now well known, were still a relatively recent discovery in the early-to-mid twentieth century.

Paul Erdös (1913–1986) was a legendarily prolific Hungarian mathematician and author of more published mathematical papers than any other mathematician. He collaborated with hundreds of mathematicians in diverse areas of mathematics, including combinatorics, number theory, classical analysis, graph theory, and probability.

Alfréd Rényi (1921–1970) was another Hungarian mathematician, a frequent collaborator and a friend of Erdös. He was primarily a probability theorist, but is also remembered for important contributions to number theory, graph theory, and combinatorics.

mathematician who is not also something of a poet will never be a complete mathematician." From the other direction, the great English poet John Dryden (1631–1700) wrote, "A man should be learned in several sciences, and should have . . . , in some measure, a mathematical mind, to be a complete poet." Of course, there are many major differences between the job of the mathematician and that of the poet. For one, poetry is in some sense purely subjective, while the mathematician is judged on grounds both objective (for example, "is this proof correct?") and subjective (for example, "are these ideas beautiful? Important?"). Let mathematician G. H. Hardy (1877–1947) have the last word here: "A mathematician, like a painter or a poet, is a maker of patterns. If his patterns are more permanent than theirs, it is because they are made with ideas."

Further Reading

James, I. M. *Remarkable Mathematicians: From Euler to von Neumann*. Cambridge, England: Cambridge University Press, 2002.

Kac, Mark, et al. *Discrete Thoughts: Essays on Mathematics, Science, and Philosophy*. Boston: Birkhäuser, 1992.

Schechter, Bruce. *My Brain Is Open: The Mathematical Journeys of Paul Erdös*. New York: Simon & Schuster, 1998.

Szpiro, George. *The Secret Life of Numbers: 50 Easy Pieces on How Mathematicians Work and Think*. Washington, DC: Joseph Henry Press, 2006.

Young, Laurence. *Mathematicians and Their Times*. Amsterdam: North-Holland Publishing, 1981.

Michael "Cap" Khoury

> ### Rota and What Mathematicians Do
>
> Mathematician and philosopher Gian-Carlo Rota wrote, "We often hear that mathematics consists mainly in 'proving theorems.' Is a writer's job mainly that of 'writing sentences'? A mathematician's work is mostly a tangle of guesswork, analogy, wishful thinking, and frustration. . . ." Rota goes on to write that the proofs emerge only later, after the mathematician has explored the problem. The proofs allow mathematicians to be sure that they are doing more than guessing and also encapsulate the relationships among different mathematical concepts and objects. It is these relationships among ideas, patterns, and structures that mathematicians are chiefly involved in exploring. Here Rota is writing against the wide gulf between what mathematicians do and what nonmathematicians typically imagine that mathematicians do. There is an also an implicit comparison of mathematicians and other professionals (such as writers) whose role also involves making meanings, exploring patterns, and explaining ideas.
>
> Gian-Carlo Rota (1932–1999) was an Italian-American mathematician and philosopher, unique in holding professorships in both subjects at the Massachusetts Institute of Technology. His mathematical work was chiefly in functional analysis and combinatorics; his main philosophical work was in phenomenology.

Mathematics, Applied

Category: Mathematics Culture and Identity.
Fields of Study: Connections; Representations.
Summary: Virtually all human pursuits depend on or were made possible by some application of mathematics, and historically applied mathematics often preceded the study of pure mathematics.

In the 1920s, the German military adapted, from a then recently developed business device, and began using an encryption code device known as the "Enigma" machine. Believing it to be an unbreakable encryption device, they continued to employ it into World War II. A significant effort to crack the Enigma code was undertaken, first by Polish mathematicians prior to the German invasion of Poland and then by British mathematicians at Bletchley Park, culminating in the breaking of the code in 1940. This success was instrumental in the ultimate Allied victory and in shortening the war significantly.

To this day, a key role of many applied mathematicians still involves encryption and cryptography, not just for military and defense purposes, but for a wide range of life activities, including computer and ATM security. In fact, more generally, there is almost no area of science, technology, and culture that is not heavily dependent upon the application of mathematical concepts and techniques. Applied mathematics therefore represents, in many ways, the ultimate multidisciplinary subject.

Historical Context

Although archaeological evidence is spotty and incomplete, it appears that the first mathematical efforts of civilized society involved either commerce, including accounting for transactions and inventories, or the measurement of land holdings for agricultural purposes. For these purposes, ancient Egyptians and Babylonians developed and applied basic concepts and techniques in arithmetic and geometry. Both peoples also used geometry in support of their building efforts and in the placement of monuments.

A large part of what is known about Egyptian mathematics comes from examination of the Rhind Papyrus. This document includes practical mathematical examples and exercises. It is apparent that the early development and purposes of mathematics were in response to, and in support of, practical, real-world problems, often of an engineering nature.

As with so many aspects of human culture and cognition, ancient Greece represented a shift—or, at least, the beginnings of a shift—in its approach to and philosophy regarding mathematics. There were certainly still applied mathematics problems, for example, involving navigation and astronomy. However, apropos of the birth in Greece of philosophical thought and reasoning, there was also some movement toward a reasoned approach to advancing mathematical knowledge. Thus, a divergence between pure and applied mathematics began to emerge.

There were several areas of mathematics in which inroads were made by the Greeks, for example, in geometry, trigonometry, logic and proof, and algebra (although work in algebra began with later Greeks). The Greeks also noted and struggled with irrational numbers (numbers that cannot be expressed as a ratio of integers). One can imagine, in an age where immediate physical needs and practical problems were paramount, that the inability to precisely measure a length (for example, not being able to precisely express the length of the diagonal of a square in terms of the known length of the sides) would have provided a conundrum.

Because of the Greek willingness to consider the theoretical, they were able to deal with, or accept, such situations to a degree. These situations led, however, to certain philosophical problems or paradoxes, such as Zeno's paradox, named for Zeno of Elea, that continue to challenge mathematicians. It can still be difficult for modern people, who initially in life are cognitively dependent upon experience and observation, to make the jump from the world of physically demonstrable, practical, applied mathematics to that of abstract and representative theory.

The Roman Empire had a largely practical, engineering-oriented approach to life, and this was manifested in their approach to mathematics. They were not particularly interested in expanding the horizons of mathematical theory. Instead, they used mathematics for applied engineering purposes from which emerged remarkable achievements that have survived through history.

A key application of mathematics beginning in the seventeenth century involved trying to understand and mathematically model the natural world. Certainly, there had been earlier efforts in that direction going back thousands of years—perhaps most notably by Ptolemy of Alexandria, with his extensive system of cycles and epicycles geared toward explaining, and ultimately predicting, the movements of heavenly bodies. But in the seventeenth century, with mathematicians and physicists such as Galileo Galilei and Isaac Newton, the modern effort to explain the world began in earnest.

Into and through the nineteenth century, a mathematician, like a scientist, was largely capable of understanding and keeping up with mathematical developments. With the explosion of mathematical activity in the twentieth century, it became impossible to do so, leading to a splitting of different specializations and mathematical disciplines and also a split between pure and applied mathematics, particularly in academic institutions. Interestingly, toward the end of the twentieth century and beginning of the twenty-first century, that separation seems to have lessened as each area began to appreciate more the usefulness of the other.

Substance of Applied Mathematics

It is difficult to comprehensively identify the substance of applied mathematics. In part, the difficulty is because of the overlap, which can take several forms, between pure and applied mathematics. First, a mathematical discovery or technique that initially seems without a practical application can, over time, become adopted and embraced by science and technology for practical application. Thus, to complain that an area of mathematics has no current usefulness can be potentially shortsighted; no one knows what future advances in society might be welcoming of—or possibly even made possible by—those pure mathematical excursions.

Another way in which pure and applied mathematics can overlap is simply in how such things are labeled. It is impossible to draw a clear line of demarcation between pure and applied mathematics. A new proof or technique made in a pure mathematics context may have very real practical applications, either now or later. Similarly, a practical, real-world problem may result in the development of a new approach with conceptual implications for theoretical mathematics. Furthermore, while many jobs require mathematical skills and techniques, such as architecture and engineering, they may not be technically classified as "applied mathematics" careers.

For example, a mathematical subject area such as number theory would generally not be considered an area of applied mathematics; and yet, it has significant implications and relevance for certain types of industrial applications, such as encoding. Similarly, abstract algebra would not, on the surface, seem to be applied; nevertheless, physicists now use group theory to better understand the world of elementary particles and quantum physics.

An important organization for applied mathematics is the Society for Industrial and Applied Mathematics (SIAM). According to its Web site, SIAM was organized in 1952 "to convey useful mathematical knowledge to other professionals who could implement mathematical theory for practical, industrial, or scientific use," and its membership in 2011 consisted of some 13,000 individuals and nearly 500 institutions.

A listing of some of the activity groups within SIAM serves to indicate the wide range of mathematics with important applications:

- Computational science and engineering
- Control and systems theory
- Dynamical systems
- Financial mathematics and engineering
- Geosciences
- Imaging science
- Life sciences
- Mathematical aspects of materials science
- Nonlinear waves and coherent structures
- Optimization
- Supercomputing

Example of an Applied Mathematics Discipline: Actuarial Science

Mathematics can be—and is—applicable to most any discipline. An example of an important and well-respected applied mathematics profession (which is generally ranked in the top five and sometimes at the very top of job-ratings surveys) is actuarial science. This applied mathematics career is representative of others and gives a sense of the general nature of applied mathematics work as well as its impact on society.

The ability to manage risk—not necessarily to eradicate or even reduce it, but at least to "manage" its potential impact—is critical in a complex socioeconomic environment. Without a way to manage risk, for example via an effective insurance industry, many activities that humans rely on might never happen (bridges might not be built and surgical procedures might not be undertaken) without the protection to society, organizations, and individuals that insurance provides. The ability to offer protection against the impact of risks is based on some key statistical ideas: the Law of Large Numbers and its related concepts. Only with a sophisticated understanding and application of probability and statistics can an effective risk management industry be sustained.

Actuarial science developed as the mathematical discipline underlying the analysis of risk contingencies. There are basically three types of actuaries: (1) life actuaries, who deal primarily with human mortality issues and life insurance; (2) pension actuaries, who focus on pension and retirement systems; and (3) property-casualty actuaries, who deal with other areas of risk and insurance, such as auto, homeowners, workers compensation, and medical malpractice insurance. Actuarial science is, in some ways, the ultimate interdisciplinary field.

Since risk applies to any type of endeavor or situation, an actuary attempting to quantify risk should potentially understand at least the fundamentals associated with almost all topics. One cannot adequately comprehend or evaluate a set of data without understanding where it came from and under what specific conditions it emerged. Thus, being an actuary or a risk analyst involves not only the relevant mathematics but also asking questions and learning about the context of the situation and using the findings to tailor mathematical methods appropriately. Furthermore, as with any quantitative discipline that uses sophisticated techniques, an effective actuary must be a very good communicator—able to translate mathematical concepts and techniques into understandable descriptions for nonmathematicians.

Becoming an actuary is a significant accomplishment. After earning an undergraduate degree (most often in either actuarial science or mathematics), actuaries spend several of their first careers both working at a job and studying for an extensive series of professional exams in an attempt to earn a designation or certification. These exams cover a variety of relevant areas, including specific actuarial techniques, finance and economics, and business processes.

On the job, actuaries use mathematics in an attempt to model real-world stochastic processes, such as the frequency and size of insurance losses, as well as economic and financial variables, such as interest rates, inflation, and investment performance. For example, based largely on historical data, an actuary might estimate that the frequency, or number, of claims that will occur in a given year is well-represented by a certain statistical distribution, such as a Poisson, named for Siméon-Denis Poisson, or a Negative Binomial. Similarly, given that a claim has occurred, historical loss information might suggest that the dollar size of a particular claim probabilistically follows another type of distribution, such as a Normal, Gamma, Lognormal, or Pareto, named for Vilfredo Pareto. Such decisions are largely based upon a thorough analysis of historical data, but other factors are also taken into account, including a qualitative understanding of the nature of the risks and hazards that the insurer is indemnifying and the entire socioeconomic context of the insurance activity. Once a model is developed, it provides a basis for not only prediction and analysis of appropriate future insurance policy rates but also testing the potential impact of making a variety of possible strategic or operational decisions, such as changes to the types of policyholders targeted and changes in policy provisions.

A well-respected applied mathematics profession is actuarial science—a truly interdisciplinary field. (Photos.com)

In the last few decades of the twentieth century, actuarial science and risk management became more technically sophisticated and more enterprise-wide in perspective. Part of the actuary's job is to understand the behavior of economic and financial variables and how they may impact the insurance and risk management process. For example, Brownian motion equations and concepts, named for Robert Brown, are frequently used to model the movements of interest rates and equity prices over time. Because insurance companies take in premiums but may not pay out corresponding losses for months or years, it is important to model how the insurer's investments may perform in the future. Insurers may even decide to sell some of their policies at an underwriting loss because they know that they have the opportunity to earn an adequate return on equity from the potential investment earnings on the premiums they take in as well as on their equity. By considering all aspects of an insurer's operations, including the effect of economic and financial conditions, the actuary's job has become much more holistic, or multidisciplinary.

Overall, an actuary's or risk analyst's job is one that is completely predicated upon mathematical techniques and quantitative skills, but it is also a business position. Skills involving communications, problem solving, project management, and teamwork are also essential for success in this environment.

Other Applied Mathematics Fields and Careers

The above description of actuarial science is representative of a variety of areas of applied mathematics. There are several other areas, including the following:

- *Biomathematics and biostatistics.* Applications of mathematics to biology have the potential to advance society and the human condition in substantial ways. Some of that advancement will come from mathematical modeling and analysis of genome and DNA mapping and sequencing. Another important area involves applying network analysis and dynamical systems techniques to the potential spread of infectious disease. Other areas include using geometry, topology, and other mathematical tools to examine and image brain activity; using differential equations and geometry to locate and attack tumors; and modeling human organs to allow testing of new surgical or other medical techniques.
- *Operations research.* Anything involving sequential processes can potentially be made more efficient and effective with the application of mathematical techniques and modeling. A few examples of such processes to be modeled include queue lines to limited resources, such as ATMs or grocery checkout machines; automobile traffic patterns; and Internet traffic. Like much of applied mathematics, the ultimate goal of operations research is to improve operational and strategic decision making.
- *Natural hazard modeling.* Hurricane and earthquake modeling are examples of interdisciplinary applied mathematics areas. Modelers need not only appropriate quantitative skills, such as geometry and systems of differential equations, but also an understanding of the appropriate science or technology, such as atmospheric sciences or geosciences.
- *Software, computers, and data.* Applied mathematics disciplines make use of computers, and some are very heavily dependent upon computational techniques and resources. In addition, numerous areas of mathematics play a role in careers in software engineering, data analysis, digitization, and compression.

Looking Forward

Ian Stewart, in the 2002 book *The Next 50 Years: Science in the First Half of the Twenty-First Century*, offers an essay titled "The Mathematics of 2050." In that chapter, he opines that several areas of mathematical exploration will undergo, and indeed are already undergoing, upswings or even revolutions. Among those he mentions are several areas of applied mathematics, including biomathematics and financial mathematics. Biomathematics certainly seems to be coming of age, and people's lives, and those of their immediate descendants, are being overwhelmingly affected by developments in this area.

One might argue, after the financial and economic crises of the first decade of the twenty-first century, that financial mathematics sustained a "black eye" that will suppress its credibility and potential. However, these same crises certainly made clear the importance of understanding the nature and potential impact of "risk" in the world, perhaps especially economic and financial risks. Being able to identify, quantify, and manage risk is critical to the smooth operation and advancement of society. This ability is simply impossible without a good understanding of the mathematical underpinnings of economics and finance and their attendant risks, as well as the ability to model different approaches and solutions to managing those risks.

It is, of course, difficult to hazard any guesses about long-term societal developments. However, one prospective application from the realm of science fiction is interesting to note. Isaac Asimov, in his *Foundation* series of stories and books, posited a mathematics-based "psychohistory." The stories focus on the legacy of Hari Seldon, a mathematician who built psychohistory into a statistical basis for modeling and predicting how human society will likely respond to various factors and stimuli. In the early twenty-first century, applied mathematicians are far from exhausting the potential of mathematics to change and advance society.

Further Reading

Holmes, Mark H. *Introduction to the Foundations of Applied Mathematics*. New York: Springer, 2009.

The Operational Research Society. http://www.orsoc.org.uk.

Society for Industrial and Applied Mathematics. "Thinking of a Career in Applied Mathematics?" http://www.siam.org/careers/thinking.php.

Tan, Soo Tang. *Applied Mathematics for the Managerial, Life, and Social Sciences.* Belmont, CA: Brooks Cole/Cengage Learning, 2008.

U. S. Department of Labor. Bureau of Labor Statistics. "Occupational Outlook Handbook 2010–11." http://www.bls.gov/oco/ocos043.htm.

Rick Gorvett

Mathematics, Defined

Category: Mathematics Culture and Identity.
Fields of Study: Communication; Connections; Reasoning and Proof; Representations.
Summary: Mathematics often begins with definitions; however, it is much more difficult to succinctly describe the whole of mathematics.

For most students, the subject of mathematics is studied on an almost daily basis from kindergarten through high school. But if asked, "What is mathematics?" the majority might struggle to formulate a meaningful answer, perhaps saying that mathematics is "arithmetic" or "algebra." But saying that the essence of mathematics is arithmetic is akin to saying that the essence of chemistry is the periodic table; mathematics and chemistry are both so much more. Not only may students have difficulty describing what mathematics is, but even professional mathematicians struggle to provide a succinct, convincing description of the nature of their subject of expertise.

In his delightful 1940 essay, *A Mathematician's Apology*, G. H. Hardy (1877–1947) spends about 150 pages presenting a passionate case for the meaning, essence, and importance of mathematics and the professional mathematician. Along the way, he offers many keen insights into the nature of mathematics:

> A mathematician, like a painter or a poet, is a maker of patterns. If his patterns are more permanent than theirs, it is because they are made with ideas.... The mathematician's patterns, like the painter's or the poet's, must be beautiful; the ideas, like the colours or the words, must fit together in a harmonious way. Beauty is the first test: there is no permanent place in the world for ugly mathematics.

This quote may appear odd; most people do not view mathematics as a creative endeavor, much less one that can be rightly considered "beautiful." Often, students who learn the subject view the discipline as one that is rigidly bound by rules, one in which there is always one right answer, and perhaps even that there is only one right path to follow. But mathematicians have a decidedly contrary viewpoint. Faced with an interesting problem to solve, the mathematician strives to have his full cadre of creativity flowing, perhaps asking: "What unusual approach might I take to solve this problem, one that nobody else has yet considered?" "How might I alter the problem to a new, related one that I might be able to solve first?" and "Is there new language or notation that I might introduce that makes the problem easier to understand or similar to another problem that is already well understood?"

More than this, as Hardy's quote alludes, mathematics is about more than individual problems; rather, it involves the study of patterns. If mathematicians can solve one particular problem, they are next interested in knowing if their methods extend to solving an entire collection of related problems. If a theorem can be proved to explain a wide class of situations, is it possible to extend the result to include even more possible scenarios? In this way, mathematics and mathematicians seek to recognize, understand, and explain patterns. Some of these patterns occur in the world around us; others may be purely theoretical. Once a pattern is understood or explained, mathematicians wonder if they have found the best explanation. What is "best"? While that is somewhat a matter of individual taste, most mathematicians agree that the best mathematics is clear, brief, and elegant. In Hardy's words, "the ideas ... must fit together in a harmonious way." It usually takes a great deal of creative insight (creative thinking, creative writing, and creative problem solving) to make the ideas fit together in a harmonious way.

Philosophers on Mathematics

Philosophers have argued for centuries, even millennia, over the nature and meaning of mathematics. There are entire schools of thought—referred to with names like "intuitionism," "logicism," and "formalism"—that seek to explain what mathematics is. However, each somehow comes up short. Perhaps mathematics itself is simply too big to describe with a formal philosophical system. Some parts of mathematics do rely on our intuition and understanding of physical happenings in the surrounding world; other aspects of the subject rely considerably on the foundations of logic; and part of mathematics grows from the formal rules that seem to many to lie at its very core. But no one of these perspectives encompasses the entire subject nor satisfactorily describes its essence. Nor does any one of these perspectives fully answer the question of where mathematics exists. Is it embedded in the surrounding world, or is it a mental construct that rightly belongs to humanity's collective brain?

Nobel prize–winning physicist Eugene Wigner (1902–1995) was one scientist who recognized the beauty, power, and harmony of mathematics while still being somewhat mystified by its nature. In his own 1960 essay on the question of what mathematics is, titled "The Unreasonable Effectiveness of Mathematics," Wigner tells the story of a statistician sharing with a friend his work in analyzing population trends. In one of his key formulas, the symbol π arises. The friend asks, "What does that symbol represent?" The statistician notes that, as usual, the symbol is the familiar π associated with circles—the ratio of a circle's circumference to its diameter. The friend is incredulous, for how can the relationship between a circle's circumference and diameter have anything to do with how a population is distributed?

This story that opens the essay illustrates Wigner's broad point: mathematics is unreasonably effective, with abstract mathematical ideas emerging in remarkable and surprising places. As a scientist trying to understand the workings of the physical universe, Wigner was particularly mystified by how well mathematics helped him describe observable phenomena. In his words, "the enormous usefulness of mathematics in the natural sciences is something bordering on the mysterious and there is no rational explanation for it." He goes on to argue that somehow the very nature of mathematics, even though it is abstract and a mental construct, leads the way in describing the surrounding world and that somehow this is indicative of a deeper truth. He concludes the essay by observing, "the miracle of the appropriateness of the language of mathematics for the formulation of the laws of physics is a wonderful gift which we neither understand nor deserve. We should be grateful for it and hope that it will remain valid in future research and that it will extend, for better or for worse, to our pleasure, even though perhaps also to our bafflement, to wide branches of learning."

To the pure mathematician, mathematics may be a quest to recognize, understand, and explain abstract patterns that arise in considered ideas. To the applied mathematician, mathematics may be a language that aptly describes patterns that emerge in some sort of physical reality. Somehow, it is the same mathematics in both cases, and the subject seems not to care whether or not it is used for abstract or applied purposes. The history of mathematics is filled with stories that show how mathematics emerges from the mental doodling of interested people, only later to find rich connections with other areas of mathematics itself, and then finally to spectacularly describe some deep physical reality.

As an example, the Greeks (c. 350 B.C.E.) came to know a beautiful number with marvelous abstract properties, today called the "golden ratio"

$$\varphi = \frac{(1+\sqrt{5})}{2}.$$

This number, approximately 8/5, arises naturally from considering line segments or rectangles that can be divided in ways that are self-similar and possesses a wide variety of interesting geometric and numeric properties. Roughly 1500 years later, people in India (c. 1150) first encountered the so-called "Fibonacci numbers": (1, 1, 2, 3, 5, 8, 13, 21, 34, . . .), which come from starting with a pair of 1s, and then adding the preceding two numbers to create the next. Spectacular patterns and relationships exist among the Fibonacci numbers, and mathematicians have been fascinated with them since. An early observation showed that the ratios of consecutive Fibonacci numbers (5/3, 8/5, 13/8, 21/13, . . .) forms a sequence of numbers that converges to

$$\varphi = \frac{(1+\sqrt{5})}{2}.$$

Much later, near the end of the twentieth century, mathematicians and biologists came to understand the apparent role that both Fibonacci numbers and the golden ratio play in explaining seed distributions in plants, such as coneflowers and sunflowers: the golden ratio appears to be the constant angle at which seeds are "born," and the relationships the golden ratio enjoys with the Fibonacci numbers help explain why this phenomena occurs, which one can better understand when the seeds in the flower are numbered.

What is Mathematics?

There is a great deal of delightful reading one can pursue to learn more about the nature of mathematics. Such investigation will help each person decide individual answers to the question "What is mathematics, really?" For the novice mathematician, Steven Strogatz has written the quintessential modern sequence of essays on the topic, essentially in the form of a blog for the *New York Times*. Strogatz, a prominent applied mathematician who has done groundbreaking work in the field of dynamical systems, begins with the wonderful 2010 essay "From Fish to Infinity," where his overall goal for the series is to be "writing about the elements of mathematics, from pre-school to grad school, for anyone out there who'd like to have a second chance at the subject—but this time from an adult perspective. It's not intended to be remedial." For the reader with a bit more mathematics background, one can consider the American Mathematical Society's Online Feature Column, a monthly column that takes a look at accessible mathematical research and (often) its applications. To begin, the interested reader might read David Austin's immensely popular 2006 explanation of Google's PageRank algorithm, "How Google Finds Your Needle in the Web's Haystack." For a more historical view, it is hard to beat the marvelous writing of Professor William Dunham in his 1990 book *Journey Through Genius*, which surveys some of the great theorems of mathematics.

An encyclopedia entry is a tiny start to describing the essence of mathematics. Each person must read, explore, think, and investigate to seek understanding of what mathematics really is. It is a beautiful example of the depth and complexity of mathematics itself that so many different perspectives on the subject ring true and that each person can find something unique in the subject.

Further Reading

Austin, D. "How Google Finds Your Needle in the Web's Haystack." *American Mathematical Society*. http://www.ams.org/featurecolumn/archive/pagerank.html.

Dunham, W. *Journey Through Genius*. Hoboken, NJ: Wiley, 1990.

Hardy, G. H. *A Mathematician's Apology*. Cambridge, England: Cambridge University Press, 1940.

Naylor, M. "Golden, $\sqrt{2}$, and π Flowers: A Spiral Story." *Mathematics Magazine* 75, no. 3 (June 2002).

Wigner, E. "The Unreasonable Effectiveness of Mathematics in the Natural Sciences." *Communications in Pure and Applied Mathematics* 13, no. I (February 1960). http://www.dartmouth.edu/~matc/MathDrama/reading/Wigner.html.

Matt Boelkins

Mathematics, Elegant

Category: Mathematics Culture and Identity.
Fields of Study: Communication; Reasoning and Proof.
Summary: A mathematical accomplishment may be considered elegant because of its conceptual depth, its aesthetic appeal, its importance and implications, its rigorousness, or the surprise of its results.

Elegant mathematics is an elusive idea, often being an aesthetical judgment determined subjectively as a reflection of one's knowledge and understanding of mathematics. That is, an intricate proof in number theory or analysis may be deemed "elegant" by mathematicians, but it would be mere nonsense to a struggling high school student. In turn, the visual "completing of the square" as a proof of the quadratic formula may be deemed elegant by high school students, but it would be too simplistic and inefficient to mathematicians. Thus, it is necessary to dig deeper into the meaning of "elegant mathematics," trying to focus on the many forms of mathematics—its methods, its visual aspects, and its role as a language.

The word "elegance" often is defined as an attribute that is effective and simple. Elegant mathematics can then be defined as mathematics that is effective and simple. However, this definition can be deceptive because one of the primary roles of mathematics is as a language, capitalizing on its ability to effectively record and model ideas and situations using a symbolic notation that is both effective and simple. Thus, since mathematics is broader than a mere language, the view of aspects of mathematics as elegant should include other attributes, such as surprise, nontriviality, consistency, power, conceptual depth, and even beauty.

When discussing elegant mathematics, mathematicians usually refer to proofs as prime examples, shifting the focus from the correctness of the proof's logical structure to its effectiveness and simplicity. Specific elements that suggest elegance are the following:

- Uses a minimal number of necessary assumptions
- Is unusually succinct yet understandable
- Avoids complex or laborious calculations
- Offers a surprising path from assumptions to conclusion
- Models "out-of-the-box" thinking
- Achieves a difficult result with a minimum of work
- Includes original conceptual insights that clarify both the "how" and the "why"
- Can be generalized to a broader context or set of problems
- Displays the power of mathematics as both a method and a language

An example of an elegant mathematical proof is Euclid's proof that an infinite number of primes exist. Using the process of "reductio ad absurdum," assume that the number of primes is finite, which may be written as $p_1, p_2, p_3, \ldots, p_n$. Let

$$N = p_1 p_2 p_3 \cdots p_{n-1} p_n + 1$$

which cannot be prime because $N > p_n$. Thus, N must be composite and have a prime factor. However, all of the known primes $p_1, p_2, p_3, \ldots, p_n$ are not factors of N because on division they leave a remainder of 1. Thus, there must exist another prime $q > p_i$ for all i, such that q is a factor of N. But, this is a contradiction of the original assumption, and the number of primes is infinite. Euclid's proof is mathematically elegant because it is effective, simple, powerful, and surprising. Many other mathematical proofs are regarded as elegant, such as these examples:

- Archimedes's use of mechanical concepts to prove that the volume of a sphere is two-thirds the volume of its circumscribing cylinder
- The Chinese "Behold!" proof of the Pythagorean Theorem
- Fourier's use of series to prove that the number e is irrational
- Euler's proofs involving infinite series
- Cantor's diagonal proof of the countability of the rationals, as well as his related proof that the reals are not countable

Paul Erdös, a Hungarian mathematician, often referred to an imaginary book in which God had included all the most beautiful or elegant proofs in mathematics. Then, when he came across a proof that he felt was elegant, Erdös would suggest, "This one's from The Book!" In the 1990s, Martin Aigner and Günter Ziegler capitalized on Erdös's ideas and published *Proofs From THE BOOK*. The most recent edition (2009) includes 30 sections involving elegant proofs from number theory, geometry, combinatorics, analysis, and graph theory.

Inelegant Proofs

A mathematics proof that is not elegant is viewed as ugly, laborious, awkward, or pedantic. Inelegant mathematical proofs often involve computer-based computations that cannot be easily replicated by mathematicians within a reasonable time frame. These inelegant yet effective proofs are akin to proofs by exhaustion involving a great number of cases, thereby disguising any elements of brevity or simplicity. A primary example of such a proof is Kenneth Appel and Wolfgang Haken's proof of the Four Color Theorem in 1976. Despite their use of some clever categorizing techniques, the final steps in the proof required more than 1000 hours of computer time to check 1,936 maps of reducible configurations as possible counterexamples. In fact, some mathematicians do not accept the proof because of its reliance on computers. Yet, the Four

Color Theorem as a conceptual statement is itself considered to be elegant.

Elegant Versus Ugly

Famous mathematicians such as Bertrand Russell, G. H. Hardy, Richard Feynman, and Paul Erdös also have shared their opinions relative to the distinctions between elegant and inelegant proofs (or mathematics, in general), often taking strong stands. For example, in a letter to Max Wertheimer, Albert Einstein even discussed the distinctions between elegant and ugly proofs. For him, a proof was ugly if it depended on the artificial introduction of additional elements, such as constructing auxiliary lines, which distracted the reader from the flow and "symmetry" of a proof. In his letter, Einstein provides both elegant and ugly examples of proofs of Menelaus's Theorem on Colinearity.

Authors have jumped on this "elegant versus ugly" bandwagon, extending it by their evaluations of both the proof and the conceptual claims associated with a mathematical theorem. The result is published resources such as *The Most Beautiful Mathematical Formulas* (1992) and *An Introduction to the World's Most Elegant Mathematics* (2006).

Unfortunately, the sorting process is not as straightforward as these authors suggest. Often, mathematicians vacillate, being unsure in the classification of a proof as either elegant or inelegant. A current example of this indecision is Andrew Wiles's proof of Fermat's Last Theorem, which conjectures that no three whole numbers a, b, and c can satisfy the equation $a^n + b^n = c^n$ for any integral value of $n > 2$. On one level, Wiles's approach was ingenious (and thereby elegant) in his use of elliptic curve theory and modular forms to solve this famous extension of the Pythagorean Theorem. And on the other hand, Wiles's final proof is inelegant because it involves more than 100 pages of very difficult mathematics that deters both mathematicians and non-mathematicians. The same can be said for the proof of the Monster Group.

Elegance

Moving the focus beyond proofs alone, mathematicians tend to classify mathematical ideas, such as theorems and concepts, as "elegant" if they establish insightful connections between two areas of mathematics that were assumed to be unrelated. The most famous example perhaps is Leonard Euler's identity that relates special mathematical constants: $e^{i\pi} + 1 = 0$.

Framed copies of this fascinating identity often will be found hanging on the walls of mathematicians' offices. It exudes simplicity and explains unexpected connections of several different mathematical ideas.

The symbolic simplicity of the above identity also illustrates the elegance of mathematics as a language. In fact, combinations of mathematical symbols with words can convey mathematical ideas that are simultaneously complex and powerful. Combined further with mathematical graphics, the elegance of mathematics as a language is enhanced by the ability to convey complex ideas efficiently, consistently, and with economy.

And, partially because of its elegance in form, mathematics also is the preferred language of the sciences and many other disciplines that involve quantitative models. Awareness of this elegance led physicist Eugene Wigner to write his famous essay, "The Unreasonable Effectiveness of Mathematics in the Natural Sciences" in 1960. He concludes his paper with the statement that "The miracle of the appropriateness of the language of mathematics for the formulation of the laws of physics is a wonderful gift which we neither understand nor deserve."

It is expected that the idea of elegant mathematics will remain an elusive one, because it is a subjective judgment of the aesthetics of proof and ideas within mathematics. Though the quandary will lead to arguments, it should not have any impact on the continuing development of mathematics. That is, because of the nature of both elegance and mathematics, it is not possible to merge them as a thinking strategy. Rather, as history has demonstrated, the mathematics is first developed and proven and only then can the aesthetic judgments (elegant versus ugly) begin. And one cannot ignore the quality of the considerable mathematics that lies between these two extremes.

Further Reading

Aigner, Martin, and Günter Ziegler. *Proofs From THE BOOK*. 4th ed. New York: Springer, 2009.

Clements, Ken, and Nerida Ellerton. "Historical Perspectives on Mathematical Elegance." *Conference Proceedings, Mathematics Education Group of Australia*, Adelaide, Australia: MERGA, 2006.

Hardy, G. H. *A Mathematician's Apology*. Cambridge, England: Cambridge University Press, 1941.

Salem, Lionel, Frédéric Testard, and Coralie Salem. *The Most Beautiful Mathematical Formulas.* Hoboken, NJ: Wiley, 1992.

Wigner, Eugene. "The Unreasonable Effectiveness of Mathematics in the Natural Sciences." *Communications in Pure and Applied Mathematics* 13, no. 1 (February 1960).

JERRY JOHNSON

Mathematics, Theoretical

Category: Mathematics Culture and Identity.
Fields of Study: Connections; Reasoning and Proof; Representations.
Summary: The complement to applied mathematics, theoretical mathematics advances the field without necessarily focusing on potential applications.

Often mathematics, as a discipline, is categorized in two general areas: theoretical mathematics and applied mathematics. When this is done, it is common to consider theoretical mathematics (or "pure" mathematics) as the part of mathematics that is carried out for the sheer pleasure of "doing mathematics," for the intrinsic beauty that lies in the study of the logical patterns and abstract relations that can be found when organizing space (geometry and topology), structures (algebra), quantities (number theory), approximations (analysis), and the thought behind these actions (logic and foundations). However, in many historical cases, the results found in theoretical mathematics have had a practical or "applied" value, often not foreseen—or even understood—until many years after their discovery. Such is the case, for example, of the non-Euclidean geometries that were seen to be coherent within their axiomatic systems but were not thought of as representing physical reality. Riemannian geometry, named for mathematician Bernhard Riemann who lived in the first part of the nineteenth century, became the mathematical context for Albert Einstein's General Theory of Relativity and led to other non-Euclidean applications in twentieth-century physics.

In much of the research done in theoretical mathematics, the focus is upon extending the field in which the particular mathematician involved is a specialist. Real world applications are not usually relevant to the activity of the pure mathematician, as these belong to the realm of applied mathematics. However, research in pure mathematics often involves the "application" of results to other mathematical objects. It is also important to emphasize that new knowledge in mathematics does not come about by experimentation but by proof.

Algebra: The Study of Structure

People often associate algebra with their experience in secondary school. Algebra studied at this level is known as "elementary algebra," and, while it is a big step in abstraction for the young student, it still focuses upon real numbers and arithmetic operations in which unknown variables are substituted for numbers. However, the abstract algebra studied and developed by theoretical mathematicians generalizes the structure of the real number system and its arithmetic operations by means of axioms and works with structures that have little to do with the numbers and operations learned in school. Some of the structures most studied in algebra include groups, rings, modules, vector spaces, and fields. These structures are defined by properties and operations. Theoretical mathematicians study the relations that are established between different representations of the same structure, or even between different structures. Once again, even though the exploration, discovery, and development of all these structures is the motivation and a goal in itself, these abstract structures have found applications in areas as diverse as crystallography, computer science, music, and physics.

Geometry and Topology: The Study of Space

The study of symmetries and rigid transformations, such as rotations, reflections, and translations, is associated with the Euclidean geometry that everyone studies in a secondary school program. Euclidean geometry arose from the need to measure and survey as territorial delimitation began to be registered and ancient civilizations developed sophisticated towns and cities. Euclidean geometry, with its study of flat space (where the shortest distance between two points is a straight line), was a faithful representation of "how the world really is." The discovery of the "other geometries" in the

1800s, when some theoretical mathematicians removed the parallel postulate from the axioms of Euclidean geometry, opened a world of possibilities (or possible geometrical worlds) for exploration on the level of theoretical mathematics. However, it was to be seen that the universe, both on the macrolevel (as in, for example, astronomy) and the microlevel (as in, for example, particle physics), could be much more faithfully described with non-Euclidean geometrical properties.

In general, geometry studies the properties that change when an object is deformed, while topology studies the properties that do not change when an object is deformed. For the topologist, a circle and a square are essentially the same, because there exists a continuous function that transforms one into the other. This is the reason that topology is often called "rubber band geometry." Whereas in geometry an object remains the same only under rigid transformations, in topology, as long as adjacent points continue to stay adjacent (which means that the object cannot be cut or twisted), the object is considered the same. The rubber band can be stretched, shaped as a square, ellipse, or circle; but points that are close remain close, and the rubber band itself does not change.

Although the study of topology is very axiomatic and theoretical, its results have had important applications in physics, biology, computer science, and robotics. For example, the study of DNA topology by applied mathematicians, together with biologists and chemists, uses results from the theoretical study of "rubber geometry."

Number Theory: The Study of Quantities

Number theory, also known as "higher arithmetic," studies the properties of the natural numbers and the integers as well as the properties of those structures that are a generalization of natural numbers and integers—those structures that maintain certain fundamental properties that these numbers possess. Some of these properties are as familiar as divisibility, prime factorization, or congruence, while others have arisen through conjectures that theoretical mathematicians have made. Some of these conjectures are extraordinarily easy to understand by any nonmathematician or young student, but they are also extraordinarily difficult to prove.

Such is the case, for example, of the now famous "Last Theorem of Fermat." The theorem states that $x^n + y^n = z^n$ can be true only for $n = 1$ or 2. For over 350 years, some of the best mathematical minds worked on this problem and could not find a proof. In 1995, a proof was presented, but it used some of the most sophisticated and modern mathematical concepts from other areas of pure mathematics to be found.

Foundations of Mathematics: The Study of Thought

The study of the foundations of mathematical knowledge includes areas such as mathematical logic, axiomatic set theory, model theory, and category theory. The quest for understanding the foundations of mathematics is also part of the philosophy of mathematics.

Much of the development of certain areas of theoretical mathematics occurred in the twentieth century; for example, topology has been based on set theory, which was presented in its axiomatic precision in the late nineteenth century. The axiomatization and actual arithmetization of infinite cardinalities was essential to much of the development of theoretical mathematics as well.

On the other hand, category theory, which abstracts the sets and functions from set theory to objects and morphisms and relies heavily on arrow diagrams to model mathematical behavior, has played an important role in pure mathematical areas, such as algebraic topology and algebraic geometry. These areas cross the rather artificial boundaries into which theoretical mathematics has been divided. For this reason, the importance of category theory can be seen, as this theory provides the notions that permit transit between different mathematical structures. At the end, theoretical mathematics is founded on the idea of the demonstration. The importance of Euclid, independently of geometry in itself, rests on the fact that he was the first to formalize the way that, to this day, theoretical mathematics is done and thought about.

Number theory has been considered by some mathematicians as a paradigm of pure mathematics. However, since the appearance of computer science, number theory has been applied in a very practical way, especially in cryptography (the encoding of information) and random number generation for statistical analysis; it has even been applied in quantum mechanics.

Analysis: The Study of Approximation

Mathematical analysis began as the process of formalization and axiomatization of calculus, whose dependence on infinitesimally small quantities that "tend" to zero did not have a rigorous foundation. Today, analysis has branched out into different areas of interest. Real analysis is the study of the properties of sequences and functions of real numbers using notions such as limit, continuity, differentiation, and integration. There is also complex analysis, which studies similar notions in the context of the complex numbers, and functional analysis, which studies these notions and others properties of functions that are seen as objects in a "function space." Probability theory is also considered an area of analysis. Indeed, probability theory is a very abstract and axiomatic subject, based on set theory and measure theory.

It is worthwhile mentioning that in the 1960s an alternative axiomatization of the infinitesimal, known as "nonstandard analysis," was developed. There are mathematicians who advocate the use of this formalization as a basis for teaching calculus, given that the concept of "limits" is often difficult to comprehend for the beginning student. As seems to be the rule in theoretical mathematics, although the mathematician does not look for applications and the main goal is to expand the particular field of study, applications of analysis have found their way into science, engineering, and economics.

The Theoretical Mathematician: Training and Workplace

The educational systems in the world are not homogeneous, although once a student is at the level of graduate studies, equivalences are usually recognized. In many countries outside of the United States, an undergraduate program consists of a complete submersion in the field and virtually no courses outside the field are taken. In the United States, the majority of fields, mathematics included, are offered as majors at the undergraduate level; therefore, the number of courses taken in the particular area is less, as there are other general education requirements that need to be fulfilled. It is also common for students to take a minor in another area or even a double major.

However, people with undergraduate degrees in any part of the world will not be formally considered theoretical mathematicians. Theoretical mathematicians will have a graduate degree, almost always a Ph.D., and graduate studies are fairly homogeneous worldwide.

People trained in mathematics will have taken a full calculus sequence (single variable and multivariable), followed by an analysis sequence. As undergraduates, they often will have taken linear algebra, abstract algebra, discrete mathematics, and usually some topology or geometry. Once a student has opted to study pure mathematics, and is in a master's program, the student will orient electives to an area of interest. At the Ph.D. level, students still have to present doctoral comprehensive exams in the subjects of analysis, algebra, and, often, topology as requisites, independently of his or her area of specialization. Students will also present a comprehensive exam in their field of interest, and then they will do doctoral research, culminating in their doctoral dissertation. There are, of course, variants to this process. Some students will specialize in several fields; some will have done research in their master;s program and have produced master's theses.

The University of Cambridge established the Sadleirian Chair in pure mathematics and, since 1863, there have been eight professors who have held it. This position is usually considered a landmark in the recognition of pure mathematics as separate from applied mathematics. Universities, in general, do not have a standard approach to the separation of theoretical and applied mathematics. Some universities have a single mathematics department; some have mathematics and statistics departments in which applied mathematics is considered a concentration in mathematics. Sometimes computer science is part of the mathematics department, although this is not common at research universities.

The natural ambience of theoretical mathematicians is academia. In this context, they can transmit their knowledge, which is the product of many years of study, reflection, discovery, and creation, to future generations. Academia is also the place where theoretical mathematicians can have the time and resources to dedicate themselves to research. There are also institutions, albeit few, that support theoretical mathematicians to do research, usually at a stage in which they have already produced results and it is clear that they have a big probability of successfully obtaining new ones.

Employment in government and industry is usually reserved for the applied mathematician. However, there are theoretical mathematicians who also have applied knowledge that makes them attractive for these positions. The theoretical mathematician who ends up in an applied context can often provide insights, because of training, that will bring about novel ways of approaching concrete problems.

It is interesting that there is very little difference in the type of work and perspectives of theoretical mathematicians worldwide. The differences have more to do with the size and extension of the university systems in different countries, but the "culture" and daily life of the pure mathematician is remarkably homogeneous.

The Germ of Theoretical Mathematics in School Mathematics

In many universities in the world, prospective schoolteachers who will be teaching mathematics must take a course, or courses, that analyze elementary mathematical concepts from an advanced point of view. This requirement is because many of the concepts that are present from the very beginning of mathematical instruction are very deep, although not necessary to understand for the young student who begins the procedure of basic mathematical operations. Felix Klein (1849–1925) is known for his work in geometry, where he demonstrated that the Euclidean and non-Euclidean geometries could be considered as special cases of projective geometry, that algebra (group theory) can be basic to the study of geometry, and other achievements in theoretical mathematics. The "Klein Bottle," a two-dimensional object from topological studies that can only be understood as a whole in a four-dimensional context, is named after him. His book *Elementary Mathematics from an Advanced Standpoint: Arithmetic, Algebra, Analysis* is made up of lectures to future teachers over a 20-year period.

At the elementary school level, for example, the concept of "infinity" is present from the moment that children learn to count with natural numbers. The notion of dimension appears when two- and three-dimensional objects are introduced geometrically, and abstraction is required when these physical objects are represented by formulas, often a first contact with algebra. The notions of number theory are omnipresent, for example, in the concept of "divisibility" and integer numbers. The concepts of "equivalence class" from algebra, and "limit" from analysis are also fundamental to work with both rational numbers and roots and real numbers and approximations. Notions from set theory and logic are implicit in teaching methods and explanations about many of the operations and concepts that are expected to be taught and learned at the school level. For this reason, the schoolteacher who is expected to communicate mathematical ideas should have a basic understanding of many of the concepts of theoretical mathematics. Further, schoolteachers who understand the broader theoretical and applied contexts of the mathematics that they teach can answer student questions and plant the seeds of ideas and connections that will later become important. It is the job of mathematicians at universities' mathematics departments to transmit these ideas to students who, while not pursuing a degree or career in pure mathematics, must understand some of its fundamental components.

Theoretical Mathematicians: Their Work and Their Views

It is usually agreed upon that until the middle 1800s, there was no clear division between theoretical and applied mathematics. Even though, arguably, Euclid's *Elements* could be considered pure mathematics, the majority of mathematicians from ancient times until the 1800s were interested in solving problems. It is also true that some of these problems, such as finding the roots of polynomials of varying degrees (which led to the development of Group Theory), might not seem to have much practical application. However, in general, mathematicians as renowned as Isaac Newton, Gottfried Leibniz, Leonhard Euler, Carl Friedrich Gauss, brothers Jacob and Johann Bernoulli, Joseph Fourier, Joseph-Louis Lagrange, Evariste Galois, or Niels Abel, in their contributions to the ideas now considered part of theoretical mathematics, usually were also involved in research in which direct applications were the central objective.

In the 1800s, the axiomatization of calculus, with its convenient but mysterious infinitesimals, was carried out by Augustin-Louis Cauchy (1789–1857) and Karl

Weierstrass (1815–1897). George Boole (1815–1864) tried to formalize the laws of thought using algebra and initiated the algebra of logic, called Boolean algebra, in which algebraic symbols represent logical forms. It is interesting that this theoretical endeavor actually laid the ground for the construction of computers and electric circuits, given that these circuits can represent complex logical operations. These mathematicians would now be considered theoretical mathematicians, as their work was oriented to expanding the mathematical areas in which they worked, not to practical applications.

Although the computer does not play the same role in the work of the theoretical mathematician as it does in that of the applied mathematician, it would be false to think that the theoretical mathematician has remained untouched by the advent of the computer. In number theory, for example, if there is a conjecture about properties of, for example, prime numbers, the numbers can be generated to billons or trillions in a short interval of time, detecting in this way if some counterexample could appear. Before this possibility arose, theoretical mathematicians could sometimes spend a lifetime trying to prove a false conjecture because it would have taken several lifetimes to generate enough numbers to arrive at the counterexample. In purely theoretical areas such as commutative algebra and algebraic geometry, computer programs have been developed that permit the calculation of, for example, Gröbner bases, named for Wolfgang Gröbner, that help to further theoretical results. The proof of the Four Color Theorem, which had been attempted by theoretical mathematicians for over 100 years, was done with the aid of the computer, which carried out the multiple calculations that would not have been possible to do by hand in a lifetime.

Of course, there are those who say this proof (of the Four Color Theorem) does not correspond to pure mathematics. This very interesting debate is a product of the transition at the beginning of the twenty-first century that coexistence with computers has become a reality. A quote from theoretical mathematician David Cox, who has played an important role in bridging this gap, is very illustrative:

> My fascination with algebra led me to algebraic geometry, which was then among the most abstract areas of pure mathematics. At the time, I never would have predicted that 25 years later I would be writing papers with computer scientists, where we use algebraic geometry and commutative algebra to solve problems in geometric modeling.

Often, quotes from actual theoretical mathematicians best give an idea of how they themselves conceive their work. These quotes may illustrate a perception of pure theoretical mathematics, perhaps not so well known to a general public, rather than absolute importance of these mathematicians over any others—an idea that will always be debatable and impossible to define:

> It is not of the essence of mathematics to be occupied with the ideas of number and quantity.
> —George Boole (1815–1864)

> No matter how correct a mathematical theorem may appear to be, one ought never to be satisfied that there was not something imperfect about it until it also gives the impression of being beautiful.
> —George Boole (1815–1864)

> Mathematics is entirely free in its development, and its concepts are only linked by the necessity of being consistent, and are co-ordinated with concepts introduced previously by means of precise definitions.
> —Georg Cantor (1845–1915)

> In mathematics the art of proposing a question must be held of higher value than solving it.
> —Georg Cantor (1845–1915)

Often theoretical mathematicians are motivated by the knowledge that their abstract research and discoveries will eventually find their way to applications in technology, medicine, or economics. Theoretical mathematics can very well be conceived of as an art by those who find aesthetic pleasure in its logic and patterns, but there is no doubt, as historically has been seen time and time again, that mathematics is science as well.

Further Reading

Cox, David. "What is the Role of Algebra in Applied Mathematics?" *Notices of the American Mathematical Society* 52, no. 10 (2005). http://www.ams.org/notices/200510/fea-cox.pdf.

"Famous Mathematics Quotes." http://www.math.okstate.edu/~wli/teach/fmq.html.

Gowers, Timothy, June Barrow-Green, and Imre Leader. *The Princeton Companion to Mathematics*. Princeton, NJ: Princeton University Press, 2008.

Klein, Felix. *Elementary Mathematics From an Advanced Point of View*. New York: Dover, 2004.

Stewart, Ian. *Concepts of Modern Mathematics*. New York: Dover, 1995.

U. S. Department of Labor. Bureau of Labor Statistics. "Occupational Outlook Handbook: Mathematicians." http://www.bls.gov/oco/ocos043.htm.

Mariana Montiel

Mathematics, Utility of

Category: Mathematics Culture and Identity.
Fields of Study: Communication; Connections; Representations.
Summary: Though doing mathematics does not necessarily require utility as an outcome, there are many examples of applications in various fields.

To discuss the utility of mathematics, there must be some agreement on the definition of the term "mathematics." Many may agree that mathematics is a pure creation of the human mind. It is a body of knowledge at which one arrives by pure reason and does not rely upon any observations of the phenomenal world. This characteristic makes it free from the limitations imposed by the particular way that human minds create experience from their understanding of the underlying phenomena. The argument comes down to the following: is mathematics the complete construction of the human mind or is it universally inherent, only being discovered/uncovered by mathematicians? Many books have been written to discuss this question, and no decision has been (or will be) made on one side or the other.

There are examples of people in the mathematics community, such as G. H. Hardy in *A Mathematician's Apology*, who see a difference between pure and applied mathematics based solely on utility and revel in the fact that nothing that they will do will be useful to humanity. This statement was in part a response to the work of Andrew Littlewood and a group of mathematicians who worked strenuously for the British War Department during World War I.

There are ample examples in the historical record of mathematics, done for its own sake, that were later discovered to be applicable to real-world problems. The theory of tensors by Giovanni Ricci-Curbastro and Tullio Levi-Civita proved to be a cornerstone for some of Albert Einstein's work on relativity. The purely algebraic area of twistor theory in physics, which predicted the existence of certain subatomic particles in the 1980s, started in the area of finite algebraic geometry. Areas that in the 1980s and 1990s were considered pure mathematics now find themselves at the forefront of application: algebraic topology used to study distribution of sensors; hyperbolic geometry used to study the extent and reach of the Internet; number theory used in architecture and cryptography; and category theory used in studying social behavior. Where will the applicable mathematics of the mid-twenty-first century come from?

What sort of mathematics should be taught in schools? This question dates back to at least the early 1800s in the United States and is discussed in the work of Charles Davies. E. R. Hedrick again raised the same question in his address to the New York Section of the Mathematical Association of America in 1933. The question arises with each new generation. Should only the mathematics that is currently known to be applicable be taught to those who will only use the tool, or should they be exposed to the whole of mathematics?

The purpose of asking this question lies in the purpose of mathematics. Should mathematics, or is mathematics, done only for its own sake? If that is the case, then why has mathematics been so useful to science? This is the question raised by Eugene Wigner in his 1960 work *The Unreasonable Effectiveness of Mathematics in the Natural Sciences*. The question and his answer have again brought to the fore this long standing argument.

Historical Context

From the earliest recordings in Babylonian and Egyptian mathematics, historians and archaeologists find problem books with problems created to train the mathematical neophytes—possibly young priests—in the algorithms that were used for building, surveying, and the like. The problems were not all applied problems but did include some examples of mathematics being done for mathematics' sake. This was not the rule, though. Most mathematics of these earlier eras seemed to have been for inherently practical purposes.

From the Western perspective, it was the Greeks under the Pythagoreans who took the idea of mathematics and made it deified. The question of the "utility of mathematics" does not skip the Platonic school. There is a quote, ascribed to Euclid in Stobaeus' *Extracts* "A youth who had begun to read geometry with Euclid, when he had learnt the first proposition, inquired, 'What do I get by learning these things?' So Euclid called a slave and said 'Give him three pence, since he must make a gain out of what he learns.'" Already the teacher has to answer the long-asked question, "What is this good for?" The Platonic school may have been one of the first in which mathematics was studied for its own beauty and internal structure—not being required to have any other purpose. Archimedes saw the utility of mathematics; whether he held the same philosophical beliefs as did the Platonists, we cannot be certain.

The Romans were extremely interested in the utility of mathematics to warfare, navigation, and architecture. It was the Greeks and the Alexandrians, though, that kept mathematics moving forward until it was rescued from the fate of much of the ancient world's science by the Islamic mathematicians. Not only did they need mathematics for navigation and geometry, but they also imbued into the geometry the need to glorify Allah with the perfectness of the geometric form.

In the Renaissance in the late thirteenth century, the early scientist, Roger Bacon, made statements about the utility of mathematics, "Mathematics is the door and key to the sciences," and "... mathematics is absolutely necessary and useful to the other sciences."

Perhaps the best summary can be found in various quotes:

> The Universe is a grand book which cannot be read until one first learns to comprehend the language and become familiar with the characters in which it is composed. It is written in the language of mathematics. . . .
> — Galileo Galilei (1564–1642)

> Mathematics is a game played according to certain rules with meaningless marks on paper.
> — David Hilbert (1862–1943)

> (Cantor's work on set theory) . . . the finest product of mathematical genius and one of the supreme achievements of purely intellectual human activity."
> — David Hilbert (1862–1943)

> From the intrinsic evidence of his creation, the Great Architect of the Universe now begins to appear as a pure mathematician.
> — James Hopwood Jeans (1877–1946)

> I have never done anything "useful." No discovery of mine has made, or is likely to make, directly or indirectly, for good or ill, the least difference to the amenity of the world.
> — G. H. Hardy (1877–1947)

> . . . enigma that researchers of all times have worried so much about. How is it possible that mathematics, a product of human thinking independent of any experience, so excellently fits the objects of physical reality?
> — Albert Einstein (1879–1955)

> As far as the propositions of mathematics refer to reality, they are not certain; and as far as they are certain, the do not refer to reality.
> — Albert Einstein (1879–1955)

> The unreasonable effectiveness of mathematics in the natural sciences . . . that the enormous usefulness of mathematics in the natural sciences is something bordering on the mysterious and that there is no rational explanation for it.
> — Eugene Wigner (1902–1995)

> [enumerating cases where structures needed in physics have already been found and developed by mathematicians] . . . long before any thought of physical application arose. It is positively spooky how the physicist finds the mathematician has been there before him or her.
> — Steven Weinberg (1933–)

> This universality of application [of mathematics] can be traced back to the fact that all aspects of Nature and areas of life are governed by the same principles of order and intelligence that have been discovered subjectively by mathematicians by

referring back to the principles of intelligence in their own consciousness.

— Maharishi Mahesh Yogi (1914–2008)

Was it not the Pisan scientist who maintained that God wrote the book of nature in the language of mathematics? Yet the human mind invented mathematics in order to understand creation; but if nature is really structured with a mathematical language and mathematics invented by man can manage to understand it, this demonstrates something extraordinary.

— Benedict XVI (1927–)

Further Reading

Barrow, John D. *One Hundred Essential Things You Didn't Know You Didn't Know: Math Explains Your World*. New York: W. W. Norton, 2008.

Hamming, R. W. "The Unreasonable Effectiveness of Mathematics." *The American Mathematical Monthly* 87, no. 2 (1980).

Lakoff, George, and Rafael E. Núñez. *Where Mathematics Comes From*. New York: Basic Books, 2000.

Russell, Peter. "Spirit of Now." http://www.peterrussell.com/Reality/realityart.php .

Sarukkai, Sundar. "Revisiting the 'Unreasonable Effectiveness' of Mathematics." *Current Science* 88, no. 3 (2005).

Scheibe, Erhard. "The Role of Mathematics in Physical Science." In *The Space of Mathematics: Philosophical, Epistemological and Historical Explorations*. Edited by Javier Echeverría, et al. Berlin: De Gruyter, 1992.

Stein, James D. *How Math Explains the World: A Guide to the Power of Numbers, From Car Repair to Modern Physics*. Washington, DC: Smithsonian, 2008.

Weinberg, Steven. "Mathematics: The Unifying Thread in Science." *Notices of the American Mathematical Society* 33 (1986).

Wigner, Eugene. "The Unreasonable Effectiveness of Mathematics in the Natural Sciences." *Communications in Pure and Applied Mathematics* 13 (1960).

David C. Royster

Matrices

Category: History and Development of Curricular Concepts.
Fields of Study: Number and Operations, Algebra, Communication, Connections, Representations.
Summary: Matrices are useful for a variety of calculations and applications.

Matrices are used throughout modern mathematics and statistics and their applications in the natural and social sciences. Matrix theory and the closely related theory of vector spaces form what is now known as "linear algebra": the study of systems of linear equations and their solutions in n-dimensional space. A matrix is a rectangular array of numbers representing the coefficients of the unknowns in a linear system. The first example of such a system and its solution using matrix operations dates from more than 2000 years ago in China. The closely related concept of "determinants" was introduced independently in Japan and Europe in the seventeenth century. The systematic development of basic matrix theory, in both its algebraic and geometric aspects, took place in the nineteenth and early twentieth centuries. This theory played a major role in the development of quantum mechanics, the branch of physics underlying many of the technological advances of the twentieth and twenty-first centuries. Matrices have been commonly explored in high school since linear algebra became a standard topic in the mathematics curriculum during the middle of the twentieth century. Contemporary applications of matrix theory are cryptography, Internet security, and Internet search engines, such as Google.

Origin of the Term

The word "matrix" comes from Latin, meaning "womb," deriving from *mater* (mother). The mathematical use was introduced by James Joseph Sylvester as "an oblong arrangement of terms consisting, suppose, of m lines and n columns, a Matrix out of which we may form various systems of determinants." At present, the word "matrix" refers to a rectangular array of numbers regarded and manipulated as a single object.

Linear Systems and Row Operations

The first calculation with such an array dates from the Han dynasty in ancient China, in *Nine Chapters of the Mathematical Art*, a practical handbook on surveying, engineering, and finance. One problem posed in the handbook is this:

> There are three types of corn, of which three bundles of the first, two of the second, and one of the third make 39 measures. Two of the first, three of the second, and one of the third make 34 measures. One of the first, two of the second, and three of the third make 26 measures. How many measures of corn are contained in one bundle of each type?

In modern notation, this becomes a system of linear equations, with unknowns x, y, and z representing the three types of corn:

$$3x + 2y + z = 39$$
$$2x + 3y + z = 34$$
$$x + 2y + 3z = 26.$$

The ancient Chinese author writes the coefficients in a rectangular array and solves the system by performing operations on this array. In modern notation, start with the 3-by-4 matrix of coefficients, and then (1) multiply row 2 by 3 and subtract 2 times row 1; multiply row 3 by 3 and subtract row 1; (2) multiply row 3 by 5 and subtract 4 times row 2:

$$\begin{bmatrix} 3 & 2 & 1 & | & 39 \\ 2 & 3 & 1 & | & 34 \\ 1 & 2 & 3 & | & 26 \end{bmatrix} \xrightarrow{(1)} \begin{bmatrix} 3 & 2 & 1 & | & 39 \\ 0 & 5 & 1 & | & 24 \\ 0 & 4 & 8 & | & 39 \end{bmatrix}$$

$$\xrightarrow{(2)} \begin{bmatrix} 3 & 2 & 1 & | & 39 \\ 0 & 5 & 1 & | & 24 \\ 0 & 0 & 36 & | & 99 \end{bmatrix}$$

The third row of the last array represents the equation $36z = 99$, giving $z = 11/4$. The second row represents $5y + z = 24$, giving $y = 17/4$. The first row represents $3x + 2y + z = 39$, giving $x = 37/4$.

Gaussian and Gauss–Jordan Elimination

This simplification of linear equations by using one variable to cancel another is called "Gaussian elimination." Carl Friedrich Gauss used it systematically in the early nineteenth century in his study of the orbit of the asteroid Pallas. An even more reduced version of a system, called "row echelon form" or "Gauss–Jordan form," was first published in a handbook on geodesy written by Wilhelm Jordan. At that time, elimination methods were considered a tool for geodesy instead of a part of mathematics.

Other Historical Developments

The closely related concept of determinants originated during the late seventeenth century simultaneously in work of Seki Kowa, in Japan, and Gottfried Leibniz, in Germany. From the modern point of view, the determinant is a function of a matrix, so it is remarkable that the study of determinants originated more than a century before the study of matrices. A systematic theory of matrices, determinants, and systems of linear equations was developed by European mathematicians during the nineteenth century: the most important contributors were Augustin Cauchy, Arthur Cayley, Ferdinand Eisenstein, Ferdinand Frobenius, Charles Hermite, Edmond Laguerre, and Karl Weierstrass. Thomas Hawkins, a historian of mathematics, who has done much research on these developments, argues that the most important motivation for this development was the Cayley–Hermite problem of determining all linear substitutions of the variables of a quadratic form, which leave the form invariant.

A famous memoir by Cayley introduced the single-letter notation for matrices together with the operations of matrix addition and multiplication and clarifies the relation between matrices and systems of linear equations: "a set of quantities arranged in the form of a square, for example,

$$\begin{bmatrix} a & a' & a'' \\ b & b' & b'' \\ c & c' & c'' \end{bmatrix}$$

is said to be a matrix. The notion of such a matrix arises naturally from an abbreviated notation for a set of linear equations, viz. the equations

$$X = ax + by + cz$$
$$Y = a'x + b'y + c'z$$
$$Z = a''x + b''y + c''z.$$

Matrix Theory

In the twentieth century, Olga Taussky-Todd became what she later referred to as "a torchbearer for matrix theory." During World War II, she worked on 6-by-6 matrices related to the flutter analysis of aircraft. She used a theorem by Russian mathematician Semyon Aranovich Gershgorin to simplify the amount of calculation and computations. The theory of solving matrix systems continues in the early twenty-first century as numerical analysts search for efficient algorithms. In addition to Taussky-Todd's own theoretical and applied work in the area, she encouraged others to join in its development. Eventually, partly because of her influence, matrix theory became a true branch of mathematics instead of just a tool for applications.

Contemporary Applications

The theory of matrices is an essential part of linear algebra, which is a highly developed branch of mathematics, with many applications to the natural and social sciences. For example, matrix mechanics, the first definition of quantum mechanics, led to the study of infinitely large matrices. Matrices also represent digital images on a computer, and, in musical set theory, matrices are used to analyze or to create compositions. Matrices containing entries other than numbers and the calculus of matrices have found importance in statistics and engineering. Typical applications discussed in modern linear algebra textbooks are network flow, electrical resistance, chemical reactions, economic models, dynamical systems, vector geometry, computer graphics, least squares approximation, correlation and variance, optimization, and cryptography.

Further Reading

Austin, David. "How Google Finds Your Needle in the Web's Haystack." http://www.ams.org/samplings/feature-column/fcarc-pagerank.

Bernstein, Dennis S. *Matrix Mathematics: Theory, Facts, and Formulas*. 2nd ed. Princeton, NJ: Princeton University Press, 2009.

Grattan-Guinness, Ivor, and Walter Ledermann. "Matrix Theory." In *Companion Encyclopedia of the History and Philosophy of the Mathematical Sciences*. Edited by I. Grattan-Guinness. Baltimore, MD: Johns Hopkins University Press, 2003.

Hawkins, Thomas. "Another Look at Cayley and the Theory of Matrices." *Archive internationales d'histoire des sciences* 26 (1977).

Katz, Victor. "Historical Ideas in Teaching Linear Algebra." In *Learn from the Masters!* Edited by Frank Swetz, John Fauvel, Otto Bekken, Bengt Johansson, and Victor Katz. Washington, DC: Mathematical Association of America, 1995.

MacTutor History of Mathematics Archive. "Matrices and Determinants." http://www-history.mcs.st-and.ac.uk/HistTopics/Matrices_and_determinants.html.

MacTutor History of Mathematics Archive. "Nine Chapters on the Mathematical Art." http://www-history.mcs.st-and.ac.uk/HistTopics/Nine_chapters.html.

Murray R. Bremner

Measurement, Systems of

Category: History and Development of Curricular Concepts.
Fields of Study: Communication; Connections; Measurement.
Summary: Various systems of measurement have been used and debated throughout history, with accuracy and precision becoming increasingly important.

Some define "measurement" as the process of determining the magnitude of a quantity. The word comes from ancient Greek *metron*, meaning "proportion," but the process itself is as old as mankind. Long before humans used calculus or algebra, they were measuring length, area, volume, time, and mass. Measures were needed to make furniture, buildings, and ritual places or in landscaping, time keeping, and making skycharts and calendars. Evidence of standardized systems of weights and measures dating back to approximately 3000 B.C.E. has been found, showing that, though measurement systems have become more refined over the centuries, the concept of measurement is an ancient one. By 1600 B.C.E., people offered silver and gold sticks in exchange for products, and in this manner money became a way to measure value.

Mathematicians generalized the notion of measurement in a number of ways, such as in length spaces, metric spaces, and in the field of measure theory. Numerous mathematicians have measures named after them, like the Lebesgue measure, named for Henri Lebesgue, and the Borel measure, named for Emile Borel. Scientists and mathematicians developed systems of measurement in order to quantify objects that were once thought impossible to measure. For instance, they have developed measurements for infinite sets, ways to measure π that are accurate to huge numbers of decimals, measures for hyperbolic geometry in which the Pythagorean theorem no longer holds, tiny-scale measurements on the quantum level or in nanotechnology, large-scale measurements of the universe, and even measurements of political opinion. Some of these measurements remain controversial, like how to assess educational achievement. Mathematicians and statisticians continue to design, refine, and improve them. In twenty-first-century mathematics classrooms, students in prekindergarten through college investigate measurable attributes, including formulas, models, and processes as well as units and systems of measurement.

Early Measurement Systems

Historical records indicate that the concept of measurement was vital for ancient civilizations, as humans needed to build dwellings, make clothing, and barter for goods. Historical study has indicated that the people of ancient Egypt, Mesopotamia, and the Indus Valley all developed systems of measurement, some of which were remarkably precise. For example, the Indus Valley people used measurements of length where the smallest division was approximately equal to 1/16 inch, as well as "yard sticks" that were exactly 33 inches in length. These measurements, while ancient in origin, were used in traditional Indian architecture and remain in use in the twenty-first century. Ancient humans typically used body parts as instruments for measuring length. The most standard unit of length that developed from ancient cultures is the cubit. The cubit was commonly defined as the length of the forearm from the elbow to the tip of the middle finger.

However, ancient Egyptian culture also defined the Sacred Cubit, which was a common cubit plus an extra hand span. The Sacred Cubit was used for constructing buildings and monuments in ancient Egypt and for surveying land. As ancient civilizations progressed and trade became more vital, standardization of measurement systems became more of a concern. Ancient peoples attempted to solve this problem by creating a rod or bar of a given length (usually a cubit) that was designated as the standard unit of measure. The rod was usually normed on a ruler's dimensions. The original rod was typically kept in a temple or other safe place, and other identical rods were created and distributed throughout the community. The number of seeds that filled a clay jar or gourd served as a measure for volume. Some civilizations also used water instead of grain. Later, stones or sometimes lumps of metal of a certain weight were used for larger units. Like the rods used for length, these stones or lumps of metal were typically kept in temples or other safe places as the official standard of weight. However, duplication of the weight provided opportunities for the deception of customers, as it was fairly easy for merchants to remove weight from a lump of metal. Therefore, inspections of weight measures became common practice, and this practice still continues through the twenty-first century. Some current forms of measurement, such as the carat, were developed out of this ancient tradition.

Standardized Measurement Systems

The English system of measurement was developed from the systems of a variety of cultures, including Babylonian, Egyptian, and Roman. From the Roman culture came the use of the base 12 system (for example, 12 inches in one foot); studies in the etymology of the English measurement units show strong Roman influence. The English system was widely used through the nineteenth century because of royal edicts that helped standardize measurements. For example, King Henry I issued a decree that the distance from the tip of his nose to the end of his outstretched thumb should be designated as one yard. This standardization made the English system very popular in various parts of the world. However, not all areas of the world recognized and utilized the English system, which motivated some to call for a single worldwide standardized system of measurement.

The idea of a single worldwide system of measurement is generally credited to Gabriel Mouton (1670). While several proposals of how such a system might be established were presented at the time, Mouton's proposal used a decimal system based on the length of one minute of arc of a great circle of the Earth. Gottfried Leibniz proposed a similar system in 1673, leading to the concept of a seconds pendulum. However, little was done for more than 100 years to further establish this system of measurement.

The metric system as it is known in the twenty-first century has its origins in the French Revolution, as the National Assembly of France commissioned the French Academy of Sciences to develop a standard of measures and weights. The system that was created was based on establishing a portion of the Earth's circumference as the unit of length. This unit of length was designated a "meter," derived from the Greek word for "a measure." Units of volume and mass were derived from the basic unit of length. The unit of mass, the gram, was found by examining the mass of one cubic centimeter of water at its temperature of maximum density. The unit of volume, the liter, was designated as the amount of water in a cubic decimeter (a cube 10 centimeters on each side). What made the metric system unique was the integral relationship between units of length, mass, and volume. Additionally, the metric system was based on the concept that smaller and larger increments were created by multiplying or dividing the basic units by powers of 10. Working with a base-10 system made the metric system easy to use, as previous measurement systems had used base-12 or base-16 systems.

The first countries that actually used the new system were Belgium, the Netherlands, and Luxembourg, around 1820. France made its use mandatory in 1840. Additionally, the metric system quickly became the standard for scientific and engineering work, which increased its use throughout the world as nations developed technologically. In 1875, 17 countries signed the "Treaty of the Meter," which officially established the standards for metric length and mass. In addition, this agreement established mechanisms for recommending and adopting refinements to the system. The metric system was officially accepted by 35 nations by 1900 and is the standard system of measurement in most nations at the start of the twenty-first century. However, traditional units are still used worldwide and conversion tables and programs ensure successful calculation. Confusion between the two systems can lead to devastating consequences, such as the 1999 crash of the Mars Climate Orbiter, which NASA attributed to the failure to convert from English to metric values.

Modern-day developments in measurement have focused on developing more precise measurement

Ancient Egyptian art shows a figure measuring weight using a scale. Volume was measured in ancient societies by determining the number of seeds that filled a clay jar or gourd. (Photos.com)

units within the metric system. In 1960, the original nations of the "Treaty of the Meter" convened as the General Conference on Weights and Measures to develop a revision and simplification of the system. From this convention, seven units of measure were established as the base units of the system: meter for length, kilogram for mass, second for time, ampere for electric current, Kelvin for thermodynamic temperature, mole for substance, and candela for luminous intensity. Also from this convention came the name of Système International d'Unités (SI), or International Systems of Units, prompting the international abbreviation of SI for the metric system.

Since that time, the General Conference on Weights and Measures has continued to develop more precise and more easily reproducible definitions of measurement units. For instance, the meter was originally a fraction of the distance from the equator to the North Pole, but this measurement was complicated by the fact that the Earth is not a perfect sphere. The late-eighteenth-century expedition of mathematician and astronomer Jean-Baptiste-Joseph Delambre and surveyer Pierre Méchain to calculate the geodesic measurement was fraught with difficulties. Later, a standardized platinum and iridium meter bar was used. In 1983, the meter was redefined to relate to the speed of light, and it became defined as the distance light travels in a vacuum in 1/299,792,458 seconds. Improvements to the metric system have been ratified by the General Conference eight times since the 1960s with the most recent taking place in 1995.

Further Reading

Alder, Ken. *The Measure of All Things: The Seven-Year Odyssey and Hidden Error that Transformed the World*. New York: Free Press, 2003.

O'Connor, J. J., and E. F. Robertson. "Mactutor History of Mathematics Archive: The History of Measurement." http://www-history.mcs.st-and.ac.uk/HistTopics/Measurement.html.

Robinson, Andrew. *The Story of Measurement*. London: Thames & Hudson, 2007.

Roche, John. *The Mathematics of Measurement: A Critical History*. New York: Springer, 1998.

CALLI A. HOLAWAY
SIMONE GYORFI

Measurements, Area

Category: History and Development of Curricular Concepts.
Fields of Study: Communication; Connections; Geometry.
Summary: Measuring area is an important mathematical calculation that has been studied for thousands of years.

"Area" is often thought of as the amount of a plane that a two-dimensional figure occupies. The name comes from the Latin word *area*, which means a vacant piece of level ground, reflecting the fact that formulas to calculate area were often developed to facilitate surveying and the calculation of the size of land plots for tax or other purposes. Some formulas to perform area calculations for simple geometric shapes were known in ancient times, while other calculations like the area of curved figures could only be approximated before the development of calculus and a greater depth of understanding regarding the constant π.

In the twenty-first century, primary school students explore attributes such as area and how it changes when the shape of an object changes. They also use formulas to find the areas of rectangles, triangles, and parallelograms and investigate the surface areas of rectangular solids. In the middle grades, the surface areas of prisms, pyramids, and cylinders are also a focus. In high school, students calculate the area and surface area of various geometric figures such as cones, spheres, and cylinders. In calculus classes, students develop a deeper understanding of area through integration techniques.

Ancient History: Egypt, Babylon, and India

The Moscow Mathematical Papyrus (c. 1850 B.C.E.) and the Ahmes, or Rhind, Papyrus (c. 1650 B.C.E.) provide evidence that Egyptians of this period had systems for calculating the areas of different geometric shapes, including triangles, rectangles, and circles. The approach to these calculations is frequently expressed using methods based on the interrelationship between different geometric figures. For instance, one problem in the Ahmes Papyrus notes that the area of a circular field of diameter 9 is the same as the area of a square field with a side of length 8. Historians of mathematics have converted these calculations to an estimate of π that is about 3 1/6 (compared to the correct value of 3.141592 . . .). No differentiation is made between exact and approximate formulas, and there is nothing resembling a proof or theorem in the modern sense—the papyrus presents ways to perform calculations.

The ancient Babylonians also had methods, preserved on clay tablets written in cuneiform, for calculating area. Historians have these tablets and inferred from the calculations values of π, such as 3 and 3.125. As with the Egyptians, methods to calculate area were often expressed by the relation between different geometric figures, and there is no evidence of proofs.

Practical needs also motivated the mathematics presented in the *Sulbasutras*, appendices to the *Vedas* (Hindu scriptures), which explain how to construct sacrificial altars. These scriptures include methods for constructing circles from squares and vice versa, indicating 577/408 as an approximation of $\sqrt{2}$.

Ancient Greeks

The ancient Greeks were able to estimate or derive many areas, in some cases building upon earlier work done by people from other cultures and civilizations. Antiphon the Sophist (480–411 B.C.E.), Eudoxus of Cnidus (408–355 B.C.E.), Archimedes of Syracuse (287–212 B.C.E.), and others approximated the area of figures like the circle by using inscribed and circumscribed polygons, a technique referred to as the "method of exhaustion." Using polygons, Archimedes was also able to show that the surface area of a sphere is four times the area of "the greatest circle in it." In mathematics classrooms, students may discover this relationship by peeling an orange and fitting the peel pieces into four circles that have the same diameter as the equator of the orange. Expressed in modern terminology, each circle has area πr^2 where r is the radius, so the surface area is $4\pi r^2$. In ancient Greece, the Pythagorean theorem, named for Pythagoras of Samos, was expressed in terms of areas of squares rather than the lengths of the sides of a right triangle; the square figure on the hypotenuse had the same area as the sum of the other two squares. Euclid of Alexandria (325–265 B.C.E.) collected the theorems of Pythagoras and other predecessors into the treatise now known as the *Elements*, which has proven to be one of the most influential mathematical textbooks in history. Heron of Alexandria (10–70 C.E.) published the *Metrica*, a treatise that collected formulas for calculating the area and volume of many different geometric figures and also presented what is known today as "Heron's formula" for expressing the area of a triangle in terms of its sides:

$$K = \sqrt{s(s-a)(s-b)(s-c)}$$

where K is the area of the triangle, a, b, and c are the length of the sides, and s is the semiperimeter (half the sum of the lengths of the sides).

Seventeenth Century

In the seventeenth century, German astrologer and mathematician Johannes Kepler applied the ideas of calculus to compute the area and volume of conic sections and casks. Reportedly, his interest was sparked while at his own wedding reception; Kepler became interested in finding a method for calculating the volume of wine in barrels that were not perfect cylinders (they were wider in the middle than at the top and bottom), meaning that the simple formula for the volume of a cylinder could not be applied. Development of differential and integral calculus, necessary to find the area of curved figures, is attributed to both German mathematician Gottfried Leibniz and English mathematician Sir Isaac Newton, in the seventeenth century. Also in the seventeenth century, the French mathematician Albert Girard published a treatise that was the first to use the abbreviations "sin," "cos," and "tan" for the sine, cosine, and tangent and demonstrated that the area of a spherical triangle depends on its interior angles, which is known as "Girard's Theorem."

Recent Developments

In the nineteenth century, the ancient area problem of squaring the circle was finally resolved. The challenge had been to construct a square that had the same area as a circle using only a ruler and compass. In 1882, German mathematician Ferdinand von Lindemann proved that π is a transcendental number, meaning that it is not equal to any finite sequence of algebraic operations on integers. This characteristic also meant that ruler-and-compass methods of constructing a square with area the same as the area of a circle of radius 1 were also doomed to failure. However, methods other than ruler and compass constructions can be used to obtain such a square.

In 1917, the Japanese mathematician Soichi Kakeya posed what is known as the "Kakeya problem," which asks whether there is a minimum region in a plane in which a needle (line segment) can be freely rotated. This area minimization problem was solved in 1927 by Russian mathematician Abram Samoilovitch Besicovitch and also in 1928 by German mathematician Oskar Perron. In the 1930s, American mathematician Jesse Douglas and Hungarian Tibor Rado published solutions to "Plateau's Problem," which requires finding the area of a minimal surface bounded by a curve. The problem is named for nineteenth-century Belgian physicist Joseph Plateau, although it was first posed in the eighteenth century by Joseph-Louis Lagrange.

Further Reading
Boyer, Carl. B. *A History of Mathematics*. 2nd ed. Rev. Uta C. Merzbach. Hoboken, NJ: Wiley, 1991.
Darling, David. *The Universal Book of Mathematics: From Abracadabra to Zeno's Paradoxes*. Hoboken, NJ: Wiley, 2004.
National Council of Teachers of Mathematics. "Principles & Standards for School Mathematics: Higher Standards for Our Students . . . Higher Standards for Ourselves." http://standards.nctm.org.
Washington State Department of Transportation. "The Metrics International System of Units." http://www.wsdot.wa.gov/reference/metrics/factors.htm.
Zebrowski, Ernest, Jr. *A History of the Circle: Mathematical and the Physical Universe*. New Brunswick, NJ: Rutgers University Press, 1999.

Sarah Boslaugh

Measurements, Length

Category: History and Development of Curricular Concepts.
Fields of Study: Communication; Connections; Measurement; Number and Operations; Representations.
Summary: The challenge of measuring lengths spurred numerous mathematical developments.

The origin of length measurements certainly predates any recorded history. One can imagine a hunter in the Pleistocene making arrows whose length only marginally exceeds the draw length of his bow, or perhaps measuring spear-throwing distance so that when hunting, throws are not wasted on animals out of range. The introduction of new technologies invariably increased the demand on the range and precision of measuring abilities. To build a house, beams need to be cut to specific lengths and notched at nearly exact positions. To build a cart or any wheeled object, lengths need to be gauged with remarkable precision in order for the wheel to have the freedom to rotate while still having weight-bearing support in all directions.

As older technologies were improved and new inventions arose, the terms and mathematics of length measurement were forced to keep pace. In order to convey the perception of length without having to give an example—indicating, for example, the width of a farming field to a friend, or the height of a horse to a potential buyer—it quickly became useful to adopt certain units (agreed-upon conventions for fixed lengths that could be used for reference when desired). Many of these units originated from roughly constant measurements of parts of the body, simply because this turned every person into a walking measuring stick. The foot and the hand are perhaps the most obvious examples of body-related units of measure. In fact, the width of the palm of the hand is roughly 4 inches (including the thumb when closed against the palm), and is still used today to indicate the height of horses. The inch was originally the width of a thumb. The cubit was perhaps the first standardized unit of length and is defined to be the length from the elbow to the tip of ones longest finger. There is some evidence indicating that the yard was defined by King Henry I to be the distance from the tip of the king's nose to the end of his outstretched thumb.

History of Standardized Measures
The adoption of widespread and official standardization began, as far as is known, in Europe during the reign of Richard the Lion-Hearted in the late twelfth century. At this time it was decreed that, "Throughout the realm there shall be the same yard of the same size and it should be of iron." During the reign of Edward I, in the late thirteenth century, additional terms were created:

> It is remembered that the Iron Ulna of our Lord the King contains three feet and no more; and the foot must contain 12 inches, measured by the correct measure of this kind of ulna; that is to say,

one thirty-sixth part [of] the said ulna makes one inch, neither more nor less.... It is ordained that three grains of barley, dry and round make an inch, twelve inches make a foot; three feet make an ulna; five and a half ulna makes a perch (rod); and forty perches in length and four perches in breadth make an acre.

This quest for standardization lasted through multiple revisions of terms and new techniques for representing the meter or the yard. In fact, measurements of weight and time evolved in very similar ways with similar revisions. These efforts occasionally reached giant proportions. In 1791, after a protracted debate over the most natural and elegant way to define these units of length, the French National Assembly decided that the meter should be defined as one ten-millionth of one-quarter of the circumference of the Earth. Using geometric techniques, they had already been able to estimate this distance to be very similar to the previously held definition of the meter. France then sent surveyors all over the globe to more exactly measure this distance. Although the surveyors often encountered hostility, occasionally being arrested as French spies, in 1799 the project was completed and a platinum bar representing the definition of the meter was created and stored in a safe location.

As technology improved, so did the definitions of units of length. In the mid-twentieth century, the meter was redefined using the wavelength of light emitted by fluorescing krypton atoms. This definition, although much more complicated, had the enormous advantage that meter could now be reproduced almost exactly by any laboratory that had sufficiently advanced equipment. No longer was the definition for the meter something that lived in isolation, requiring careful guarding. Once the laser was invented in 1960, it became practical to redefine the meter in terms of the speed of light, often considered the ultimate physical constant. Thus the meter became, precisely, the distance traveled by light in a vacuum during 1/299,792,458 seconds—a definition that continues to be used at the beginning of the twenty-first century. This definition, of course, gives rise to the question of exactly how a second is defined.

Other Considerations

During this time, however, there were many more complicated considerations than simply how to define a unit of length. Once the units of length were defined, it was invaluable to have the ability to calculate the lengths of objects that seemed difficult to measure or to predict the lengths of objects that did not yet exist. When building a house, the builder must decide first how wide and how deep and how high the house is to be, and then the builder will cut down trees of the right size, and trim them down to obtain the needed logs. But how can the builder be sure that a tree is of the right size? If a builder is planning to cut down a tall tree, it is usually impractical to climb it just for the sake of measurement. Because of this impracticality, people adopted several clever methods for estimating the height of a tree without having to leave the ground.

Native Americans had a particularly clever tool-free method. They would bend over and look through their

Once units of length were defined, humans gained the ability to calculate the lengths of objects that seemed difficult to measure. (Photos.com)

legs at the tree. Then they would walk away from the tree and repeat this process until they found a point where they could just barely see the top of the tree. It turns out that the distance from this point to the base of the tree is almost exactly the height of the tree (provided the measurer is an average-sized person). The reason for this measurement technique is that a normal person looking between one's legs sees at about a 45-degree angle upward. Geometric principles indicate that since a tree makes roughly a 90-degree angle with the ground, then the measurer, the base of the tree, and the top of the tree make a 45-45-90 triangle. Such a triangle has equally long legs, which means that the height of the tree is equal to the distance from the base to where the measurer is standing. The Native Americans probably did not think about it in these exact terms, of course; they most likely discovered this trick by trial and error. Nonetheless, the ability to make these calculations is extremely important when it comes to building large structures.

Measuring Triangles

Ancient civilizations have long known that when building structures that need to hold weight, triangular supports are very effective. A natural question, then, is how long to make the triangular piece. Say a person is building a simple box to stand on. The box will be 1 meter wide, 1 meter deep, and 1 meter tall. If the person builds just the box, there is a danger it will collapse when stood upon, so triangular supports are included. Specifically, this person decides to build each of the four "wall" sides to be a square with a single piece added in diagonally to form two triangles. Since each of the squares is 1 meter tall and 1 meter wide, how long should the single piece of wood be so that it can join opposite edges of the square?

Pythagoras, a Greek philosopher and mathematician who lived around 500 B.C.E., developed a simple formula, the Pythagorean Theorem, that can be used to answer the question. His formula states that for a triangle where one of the angles measures to be 90 degrees, and a, b, and c are the side lengths where c represents the hypotenuse (the side across from the 90-degree angle) then $a^2 + b^2 = c^2$. Using this formula, the square can be imagined as two triangles, each of which has a 90-degree angle, and it can be seen that c^2, the length of the hypotenuse squared, must be equal to $1^2 + 1^2 = 2$.

Therefore, the length of the piece of wood needed is 1/7, which is approximately 1.4 meters.

When Pythagoras answered this question, it provided a huge boost to the ability to manufacture precisely engineered constructs. But practically and mathematically, it also raised additional questions. How can one determine the length of the triangular piece if the triangle does not possess a 90-degree angle? What are the properties of triangles that are best at supporting weight? The mathematical field of trigonometry (from the Greek words *trigonon*, meaning "triangle" and *metron*, meaning "measure") was invented to answer these questions.

Measuring Curves

The ancient Greeks were obsessed with geometry and incorporated it into many nonmathematical aspects of their lives. This incorporation led to the aesthetic, if intellectually dubious, concept of sacred geometry that infuses some spiritual movements. Perhaps the single most prevalent concept in geometry is that of a circle. The concept of relating various geometric shapes to circles can be seen anywhere, from images showing triangles (or more complicated shapes) connecting to a circle in various ways to Leonardo DaVinci's drawing of the Vitruvian Man. Although the circle relates heavily to the measurements computed via trigonometry, it is the source of a different and altogether more elusive measurement of length—the length of curves.

Of course when one thinks of the measurement of length, one generally considers the length of straight line segments. But even thousands of years ago, it was obvious to some mathematicians that it made sense to ask about the length of a curve. It is easy enough to draw a circle, take a piece of string, mold it to nearly the same shape as the circle, cut it off at the right point, and then remove the string, lay it in a straight line, and measure it. This measurement gives, roughly, what is called the "arc length" of the circle. To attempt to compute this length using pure mathematics, mathematicians would undergo a tedious process where they would approximate the curve using dozens, or hundreds, of small line segments. Then they would measure each tiny line segment and add up the results to get, again, roughly, the arc length of the curve. It was through this process that mathematicians discovered the remarkable fact that—although circles could be made with very large or very small arc lengths—for any circle, its arc length (also called "circumference" when refering to a circle) divided by its diameter was always a fixed number. This fixed number is π, which is approximately equal to 3.14. This fact was known to the ancient Egyptians, who, like the Greeks, had a penchant for incorporating mathematical references into culture, literature, and architecture. In fact, the Great Pyramid at Giza was built with a perimeter of 1760 cubits and a height of 280 cubits. 1760 ÷ 280 is almost exactly equal to 2π.

Calculus

In part because of the appeal of discoveries made about the arc length of circles and in part because of the practical application, there was more research done into calculating arc lengths of curves, and in general calculating other abstruse quantities, such as the area enclosed by a curve, or how wind would change the velocity of a balloon. The bulk of the theory necessary to make these computations was the mathematical field of calculus, coinvented by Isaac Newton and Gottfried Wilhelm Liebniz.

The fundamental concept inherent in calculus is to break up an object into a very large number of very small pieces and to put those pieces back together again. The advantage calculus has over the cumbersome approach used by earlier mathematicians (breaking up a curve into many individual line segments, measured individually) is twofold. First, instead of using large numbers of small pieces, they actually used infinitely many infinitely small pieces. This means that instead of getting only an approximation, the error was infinitely small, and so the methods of calculus would actually yield exactly the correct answer. The second advantage is that calculus incorporates many methods to simplify these calculations involving infinity. These techniques are so simple that many high school and college students routinely master the subject. However, a lingering flaw in calculus after Newton and Leibniz's development was the fact that the notion of "infinitely small" and "infinitely big" was vague and never precisely defined.

Mathematician Augustin-Louis Cauchy, in the mid-1800s, created the precise definition of these elusive concepts. An infinitely small quantity was defined to be a sequence of numbers that got arbitrarily close to zero; for example, 1, 1/2, 1/3, 1/4, 1/5, ... To make this even more precise, Cauchy pointed out that this sequence had a special property. To illustrate, pick as small a positive number as you can, for example, 1/1,000.000. Draw a circle around the point 0 of that small radius. At some point along the sequence, all the terms past that point will lie inside that circle—in other words, all the terms past that point will have distance from the point 0 less than the specified number. Even if you chose a new number, much smaller than the first one you picked, that would still be true—you would simply have to traverse farther along the sequence before you would find that special point. This property is called convergence—and clearly relies heavily on the notion of distance for its definition.

In the late 1800s and early 1900s, there was a large amount of work done in improving the techniques and perfecting the details of calculus. The mathematical operation that allowed one to find the area enclosed by a curve was called an "integral." One of the more important improvements to calculus, created by Henri Lebesgue in 1901, was the Lebesgue integral, a concept that extended and strengthened the original idea and allowed the development of more robust mathematical machinery. An interesting feature of this new integral was that it was so general it could be applied to curves that in some sense couldn't even be graphed. Soon after the development of the Lebesgue integral, a mathematician named Maurice Frechet, impressed by the generality of Lebesgue, invented metric spaces.

A metric space is a very general idea. It is the concept that one begins with a group of objects about which absolutely nothing is known, except that the distances between them are measurable. This idea turned out to be enormously powerful because it was able to capture the precise definitions made by Cauchy while at the same time being so general that they could apply to almost any mathematical system that people wished to study. The genius of the idea was in the realization that in so much of the complicated mathematics that was now being done, the one idea always relied on was that of measuring distance. Cauchy defined an infinitely small quantity to be a collection of numbers that becomes arbitrarily small. But this definition can be generalized to a collection of these objects, where the distance between them becomes arbitrarily small. Metric spaces quickly permeated all areas of mathematics, and metric space theory remains one of the foundational components of the mathematical area of analysis, the branch of mathematics used most heavily by scientists.

Further Reading

Alder, Ken. *The Measure of All Things: The Seven-Year Odyssey and Hidden Error That Transformed the World.* New York: The Free Press, 2002.

National Physics Laboratory. "History of Length Measurement." http://www.npl.co.uk/educate-explore/factsheets/history-of-length-measurement/history-of-length-measurement-%28poster%29.

University of Cambridge NRICH. "History of Measurement." http://nrich.maths.org/2434.

Whitelaw, Ian. *A Measure of All Things: The Story of Man and Measurement.* New York: St. Martin's Press, 2007.

Tristan Tager

Measurements, Volume

Category: History and Development of Curricular Concepts.
Fields of Study: Communication; Connections; Geometry; Measurement.
Summary: Volume has been measured in numerous ways throughout history, with calculus playing an integral role.

Volume is the amount of space that is occupied by an object. In other words, volume is a three-dimensional analogue of the area. Volume is important in construction, engineering, and physics. Early on, weight was easier to measure than volume, especially because crops and other real-life objects often had irregular shapes. One early volume calculation can be found in an ancient Egyptian mathematical work called the "Moscow Papyrus," named for the country where it resided in the twentieth century. It dates back to almost 2000 B.C.E., and its author is unknown.

One problem provides a method for calculating the volume of a truncated pyramid. However, mathematicians developed a variety of methods to calculate volume, including via the displacement of water, the method of exhaustion, and connections to determinant and integration methods. In twenty-first-century mathematics classrooms, students investigate volume relationships and formulas. Students calculate the volume of geometric objects like cylinders, cones, and spheres. The volume of a cylinder ($\pi r^2 h$), is obtained by multiplying the area of the base, πr^2 where r is the radius of the given circle, by the height (h). The volume of a cone is one-third that amount, and the volume of a sphere is

$$\frac{4}{3}\pi r^3.$$

However, these formulas took a long time to develop and could only be approximated before the development of calculus and a fuller understanding of π. Students also compare quantities of water or sand-filled relational geosolids and examine volume integrations and the volume interpretation of determinants.

Early Methods

The method of exhaustion has long been used to estimate volumes. Democritus of Abdera is noted as the first to state that the volume of a cone is one-third that of a cylinder of the same height and radius and that the volume of a pyramid is one-third of the corresponding prism. Eudoxus of Cnidus developed the method of exhaustion that uses what would now be referred to as "limits" of sums of well-known areas or volumes. He justified Democritus' relationships and explored other areas and volumes. Some of Eudoxus's work appears in Euclid of Alexandria's *Elements*. In ancient China, volume calculations were published in the Nine Chapters on the Mathematical Art. In his commentary of 263 C.E., Liu Hui calculated the volume of figures like a tetrahedron and the frustum of a cone. The volume of the sphere was challenging and he noted: "Let us leave the problem to whoever can tell the truth." Archimedes of Syracuse researched the volume of various figures, including surfaces of revolution. He showed that the volume of a cylinder equals the sum of the volume of a cone of the same height and the volume of a sphere of the same height. Here, the height can be expressed as twice the radius. Archimedes is reported to have considered this as one of his greatest achievements because a related inscription appeared on his tombstone. In twenty-first-century classrooms, students understand Archimedes' statement by pouring sand or water from a cone and a sphere into a cylinder to fill it up. They also investigate the related formulas. Archimedes also reportedly noticed that water displacement could be used to measure volume while famously expressing: "Eureka!"

Using Calculus and Integration

Computing volume by integration allowed for the calculation of the volume of irregular objects. In 1615, Johannes Kepler published *Nova Stereometria Doliorum Vinarorum* (*New Solid Geometry of a Wine Barrel*). He apparently became interested in the volume of casks on his wedding day. Methods of integration and volume calculations developed along with calculus before the related analysis was well understood. Cavalieri's principle is named for seventeenth-century mathematician Bonaventura Cavalieri. Cavalieri combined the method of exhaustion with Kepler's work and computed volumes by comparing cross-sectional areas. This method predates the analysis that was needed to put it on a sound footing, and Cavalieri was criticized for his ideas. Some results seemed counterintuitive and provided additional fodder for critics. For example, the surface of revolution obtained from revolving the region under $1/x$ between 1 and infinity has finite volume. In the seventeenth century, mathematician Thomas Hobbes is noted as having remarked about this result: "To understand this for sense it is not required that a man should be a geometrician or a logician, but that he should be mad."

Mathematicians eventually developed the analysis rigorously. The Riemann integral is named for nineteenth-century mathematician Bernhard Riemann. Measuring the area below a graph of the function is accomplished by dividing the region under the graph into extremely small rectangles and adding these rectangles up. Roughly, the volume of a region in space would be computed with a similar idea. The given space would be divided into small rectangular boxes. Each piece would have the volume $(dx \times dy \times dz)$, and the volume of the whole space would be computed by a triple integral. However, this method supposes that one understands the functions that make up the surface. Many mathematical theories about approximation of the boundary surface have been developed for a long time, and they have played an important role since they are indeed extremely useful in actual computation.

Other Methods

Another method of computing volume that is explored in linear algebra and physics classes is by the determinant. In a 1773 paper on mechanics, Joseph Lagrange calculated the volume of a tetrahedron in terms of the locations of the coordinates. In modern terms one would recognize the connection of the expression to a determinant calculation of a 3-by-3 matrix.

Integration formulas such as Green's theorem and the divergence theorem, which are studied in a multivariable calculus course, connect volume to other calculations. Mathematician and physicist George Green worked on vector calculus integral theorems, and Green's theorem is named after him. Green's theorem relates surface and volume integrals. Mathematician Carl Friedrich Gauss contributed to the geometry of surfaces as well as the divergence theorem. The divergence theory relates the volume integral of the divergence inside a surface to the flux of a vector field on the closed surface. A well-studied question related to volume measurements dates back to ancient Greece. Archimedes and Zenodorus examined the sphere as the surface that would enclose a given volume with the least amount of surface area. Mathematician Hermann Schwarz proved that the sphere maximized volume with minimal surface area in 1884. In the twentieth century, mathematicians solved the Double Bubble Problem, showing that a standard configuration is the most efficient way to enclose two regions.

Further Reading

Anastassiou, George A., and Karl-Georg Steffens. *The History of Approximation Theory: From Euler to Bernstein*. Boston: Birkhäuser, 2010.

Hirshfeld, Alan. *Eureka Man: The Life and Legacy of Archimedes*. New York: Walker, 2009.

Lawn, Richard E., and Elizabeth Prichard. *Measurement of Volume*. Cambridge, England: Royal Society of Chemistry, 2003.

Hyungryul Baik

Measures of Center

Category: History and Development of Curricular Concepts.
Fields of Study: Communication; Connections; Data Analysis and Probability; Geometry.
Summary: Mode, median, and various averages (including the arithmetic mean and weighted average) are all examples of deriving a central value.

Mathematician and anthropometry pioneer Adolphe Quetelet is sometimes called the "father of the average man." His nineteenth-century work *A Treatise on Man and the Development of His Faculties* outlined theories regarding distributions of human traits. Whereas others before him had applied the normal distribution to describe measurement errors, Quetelet asserted that human traits, both physical and intellectual, were normally distributed around some central value. In his later work, the "average man" was sometimes presented as an ideal human, a concept that mathematicians, such as Antoine Cournot, disputed. Nonetheless, notions of a central value representing a typical case within a set of observations became very influential in research and statistical data analysis. There are many ways to think about center or typical values.

One of the most common measures is the arithmetic mean, often simply known as "average," which some trace back to Pythagorean writings on properties of music. Other types of averages include harmonic, geometric, trimmed, weighted, and moving or rolling means. Measures of center besides the mean include median and mode. Statistician Frank Zizek stated in his 1913 book *Statistical Averages*:

> An average may be computed for its own sake, merely to obtain a comprehensive characteristic expression for a series of divergent values, but it is often found as a means to another end, mainly for purposes of comparison. . . .

Students in twenty-first-century classrooms may use measures of center in the primary grades, focusing on mode and median, while mean is introduced in the middle grades. Expected value, which is the long-term average for a random variable or process, is a probability concept most commonly addressed in high school or college.

Mode

In the nineteenth century, psychologist and physicist Gustav Fechner studied the nonlinear relationships between subjective psychological sensations and the actual physical intensity of different stimuli, a field now known as "psychophysics." Use of the mode as a measure of center occurs in Fechner's work, which appears to be the first mention in print. He defined it as the value "around which the items . . . collect most densely, so that equal intervals contain more items the nearer the intervals lie to this value." Later in the same century, Karl Pearson would use a probabilistic and graphical approach to the definition, stating that the mode was the "abscissa corresponding to the ordinate of maximum frequency." Consistent with the probabilistic approaches used by both Fechner and Pearson, mode has come to be defined as the most frequently occurring value in a probability distribution or set of data. It is the only measure of center that is appropriate for both categorical and numerical variables because it does not require the data to be ordered in any way.

Median

Fechner's work also contained reference to medians, which he called the "middlemost ordinate" or "centralwerth" of an ordered series of values or data points. Some credit Carl Friedrich Gauss for "inventing" the median earlier as part of his work on the normal distribution. The name "median" is attributed to Francis Galton in the late nineteenth century. Inspired by Quetelet, Galton researched ways to measure and express center and variation in data, both numerically and graphically. He devised the "ogive graph," named after a curve common in architecture and ballistics, which graphed data versus ranks. His method of "statistics by intercomparison" used quantiles and percentiles, including the median, to consider deviations. Galton's median represented a typical value, which he termed "mediocrity," often assigning it a standardized value of zero as a point of reference for comparisons. Subsequently, many nonparametric (also called "distribution-free") statistical methods based on medians were developed by mathematicians and statisticians. Some of these procedures are named for them, including Henry Mann, William Kruskal, W. Allen Wallis, Donald Whitney, and Frank Wilcoxon.

Mean

Though the exact age of either mode or median is unknown, available evidence suggests that the mean may be older. In the Pythagorean treatise, *On Music*, from the school named for Pythagoras of Samos, there is some discussion of finding the middle value of two data points, such that the value exceeds the lower value by the same amount that the upper value exceeds the middle. While this basic description could be either the mean or the median for the case of two points, some historians consider this to be evidence of Pythagorean use of the mean. Statistician and historian Robin Plackett examined evidence from Babylonia, Egypt, and Greece and concluded that, while the mean may have been used in selected cases, it did not appear to be standard practice among astronomers and others who were typically collecting data. He credits sixteenth-century astronomer Tycho Brahe with introducing the mean into scientific methods of the times.

Mathematician Thomas Simpson showed in the eighteenth century that an average was a better measure than a single observation in a very limited set of cases and astronomers often used probability and means to quantify errors of deviations in observations. Other mathematicians, such as Joseph Lagrange, Abraham de Moivre, Pierre-Simon Laplace, and Carl Friedrich Gauss, contributed to mathematical developments that addressed the mean of a probability distribution or data set in the eighteenth and nineteenth centuries, while Quetelet, Galton, and others sought novel applications of measures of center. Statisticians in the twentieth century continued work on means, including George Box and Gwilym Jenkins, whose research about moving averages is the basis for many time-series forecasting models, and new research is ongoing into the twenty-first century. The mean has many mathematical properties that make it more desirable for widespread use than the median, such as connections to the least squares criterion and the method of moments. Many statistical techniques are concerned with estimating and comparing means.

Rules and generalizations have been devised and taught over the years regarding which measure of center is best to use for any given set of data, particularly with regard to choosing between the mean and median. Mathematically, the arithmetic mean minimizes the sum of squared distances of all points from the center, while the median minimizes the sum of absolute distances. For data with perfect symmetry, these are equivalent. Data with skew or outliers may yield very different outcomes. Mode is also less clear as a measure of center if there are no repeated values or if there are two or more values that occur most frequently. In the twenty-first century, educators like sociologist and statistician Paul von Hippel continue to investigate methods to teach concepts and relationships between mean, median, mode, and skew.

Further Reading

Stigler, Stephen. *The History of Statistics: The Measurement of Uncertainty Before 1900*. Cambridge, MA: Belknap Press of Harvard University Press, 1990.

Von Hippel, Paul. "Mean, Median, and Skew: Correcting a Textbook Rule." *Journal of Statistics Education* 13, no. 2 (2005). http://www.amstat.org/publications/jse/v13n2/vonhippel.html.

Sarah J. Greenwald
Jill E. Thomley

Measuring Time

Category: Space, Time, and Distance.
Fields of Study: Measurement; Number and Operations; Representations.
Summary: A variety of mathematical calculations are used to define, measure, and apply measurements of time.

Time measurement (chronometry) serves two tasks: (1) indication of temporal instants, which are events occurring in time and (2) determination of temporal extensions (durations), which are amounts of time between events. Both types of tasks are essential for practical purposes, such as organization of life in civilized societies or intersubjective synchronization of various activities, as well as for the scientific study of nature and society. Time measurement relies upon the arithmetic model of time: events are mapped onto a numerical continuum so that if event A precedes event B, the relation $t_A < t_B$ holds between their time indices; such a mapping is called a "timescale." Given a timescale t, durations can be calculated as differences between time indices; and conversely, time indices can be defined by durations elapsed from a certain reference event (epoch).

Clocks and Timescales

Theoretical chronometry studies mathematical properties of timescales, while practical chronometry (called "horology") is concerned with devices realizing timescales, such as clocks. Any physical system, natural or artificial, producing a series of distinct and observable—thus countabl—events can serve as a clock. Periodic processes in our lifeworlds, such as the day/night cycle, the lunar cycle, or seasonal changes, provide natural bases for timekeeping and measurement. Counting recurrent observable events yields a measure of durations longer than the clock's basic period; measurement of shorter times than the clock's base period enforces a subdivision of the period into equal subunits—a refinement of the timescale. Therefore, time measurement spanning several orders of magnitude requires alignment of timescales defined by different physical processes, periods of which are in constant arithmetic relations and thus define together a common timescale. For this purpose, timescales generated by artificial clocks are used.

Predecessors of clocks were devices used to indicate one standard time period, for example, outflow water clocks (clepsydrae) or sand glasses or burning candles. A clock in the proper sense of the word generates series of events at equal periods between them. In history, different physical principles have been used to ascertain isoperiodic clock action, including mechanical oscillations of a pendulum or a balance wheel, vibrations of quartz crystals or molecules in electromagnetic fields, and electromagnetic radiation emitted/absorbed by atoms, providing high-precision frequency standards. Base periods of clocks vary from a few seconds to fractions of a second in mechanical clocks, down to an accuracy 10^{-10} seconds per day for atomic clocks, as established by national standards agencies.

The Second

The second (s) is the fundamental unit of temporal duration. Originally, the second was defined as a constant fraction (1/86,400) of the solar day. Hence its name: 1 day is 24 hours, 1 hour is 60 minutes (*pars minuta prima*, which means "first minor part"), 1 minute is 60 seconds (*pars minuta secunda*, which means "secondary minor part"). With increasing precision of time measurement, fluctuations of the Earth's rotation period had become evident: thus the second was redefined in the 1950s as a constant fraction (1/31,556,926) of the period of the Earth's orbital motion around the sun (ephemeris time). In 1967, a new definition of the second was adopted; one second is defined as a constant multiple (9,192,631,770) of the period of electromagnetic radiation emitted by cesium atoms in transition between two defined energetic states under precisely specified conditions. Thereby astronomic definitions were abandoned in favor of standards derived from the inner structure of matter, which is considered constant throughout the Universe.

Time Measurement Technologies

Advanced time measurement and synchronization technologies allow people to define a unique timescale to be used all over the world. Historically, the first universal timescale was Greenwich Mean Time (GMT), based on telescopic observations at the Royal Observatory in Greenwich, England, which was later replaced by the international Universal Time (UT). At present, the most precise basis for timekeeping is International Atomic Time (TAI), based on a worldwide network of atomic clocks. Coordinated Universal Time (UTC) is the basis for international timekeeping. UTC differs from TAI by an integer number of seconds to approximate UT. Irregularities in Earth's rotation are

A modern sundial. Predecessors of clocks also included hourglasses and burning candles. (iStockphoto)

compensated by adding or subtracting a leap second to/from the UT-TAI offset, if necessary; the corrections take place on June 30 or December 31. In this way, uniformity of UTC with respect to the atomic time unit definition is maintained, and continuity with the astronomical timescale is preserved.

Time Calculations

Irreversible natural processes, laws of which are well known, can be used to calculate time extensions, particularly those escaping direct observation and measurement. Of special importance are estimations of geological or archaeological age based on radioactive decay of certain elements or particular isotopes of otherwise stable elements. Estimates of time extensions in astrophysics are, to a large extent, theory based (for example, dependent on stars' evolution models). This dependency applies a fortiori to time magnitudes discussed in cosmology. Any time measurement implies observational (or at least conceptual) separation between the measured process and the reference process, defining the timescale. If the universe as a whole is considered, such separation is no more possible, so that the notion of the universal "cosmic" time meets logical difficulties.

Finally, there is no direct evidence that timescales defined by different classes of physical processes (inertial motions, light radiation, or radioactive decay) are really equivalent: the "unity of time" in physics is a convenient hypothesis, not an empirically secured fact. Since precise time measurements are available only for a short historical period—negligibly short relative to cosmological orders of magnitude—the alignment of radiation-based and motion-based timescales is merely temporally "local." Some cosmologists suggested that different classes of physical phenomena may define different timescales between which a nonlinear relation may hold. Consequently, two or more different time measures might be needed for adequate description of cosmic processes on a large scale.

Further Reading

Bergquist, J., S. Jefferts, and D. Wineland, "Time Measurement at the Millenium." *Physics Today* 54, no. 3 (2001).

Galison, P. *Einstein's Clocks and Poincaré's Maps: Empires of Time.* New York: Norton, 2003.

Lippincott, K., et al. *The Story of Time.* London: Merell Holberton, 1999.

Wackermann, J. "Measure of Time: A Meeting Point of Psychophysics and Fundamental Physics." *Mind and Matter* 6, no. 1 (2008).

Whitrow, G. J. *The Natural Philosophy of Time.* 2nd ed. Oxford, England: Clarendon Press, 1980.

———. *Time in History.* Oxford, England: Oxford University Press, 1988.

JIRÍ WACKERMANN

Multiplication and Division

Category: History and Development of Curricular Concepts.
Fields of Study: Algebra; Communication; Connections; Number and Operations.
Summary: Scholars throughout history have developed a variety of algorithms to compute multiplication and division.

Multiplication and division of numbers are useful for scaling a quantity, which is fundamental in any quantitative society. For example, to determine the correct cost for purchasing more than one unit of some item, the buyer should multiply the number of units by the unit price. One of the most common exposures people have to the concept of multiplication is the idea of repeated addition, which is frequently taught in elementary school. Some mathematicians object to this analogy, since it fails in specific instances, such as the case of fractions. Multiplication in this context can be thought of as a scaling process. Multiplication of mathematical objects is an operation that combines those objects in some representative manner. For example, matrix multiplication is the composition of the corresponding linear transformations.

Some theorize that the first evidence of multiplication is the Ishango Bone, a tool from Upper Paleolithic era, which may demonstrate multiplication by two. The ancient Chinese as early as the Warring States period (475–221 B.C.E.) developed a system of multiplication using a place-value system and counting boards for calculations. Multiplication using a place-value system of Hindu–Arabic numerals dates back to Indian mathematician and astronomer Brahmagupta in the seventh century.

Division is an operation that is generally the inverse of multiplication. For whole numbers, division can be thought of as finding the number of identically sized groups into which a number of individuals can be divided or partitioned. The remaining individuals, after all identically sized groups have been removed, are called the "remainder." People in ancient Egypt commonly divided food and other supplies and formalized the notion of division. Concepts of division and multiplication are generalized and studied in the fields of number theory, algebra, and numerical analysis.

History of Multiplication Algorithms

One of the earliest methods for multiplying whole numbers is called the "Russian Peasant algorithm," but a similar procedure was described thousands of years earlier in ancient Egyptian papyri and is based on doubling the multiplicand. Assume that an Egyptian scribe wanted to compute the product of 13 and 23. The scribe would compute the products $1 \times 23 = 23$, $2 \times 23 = 46$, $4 \times 23 = 92$, and $8 \times 23 = 184$. By doubling each previous result, the scribe would, realizing the sum of the multipliers 1, 4, and 8 equals the multiplier 13, find the sum $23 + 92 + 184 = 299$, the required product. Thus, multiplication is reduced to being able to both double a number and add.

Another method used to multiply numbers is called "gelosia" or "lattice multiplication," which is still taught in some elementary schools. This method probably originated in India before the twelfth century and eventually became the inspiration for Napier's bones, an instrument created by John Napier, which was used to accomplish multiplication after its invention in the early 1600s. To multiply, for example, 23×48, one draws a 2-by-2 lattice of squares where each square is bisected by a diagonal from upper right to lower left. The rows are labeled, top to bottom, 4 and 8, while the columns are labeled, left to right, 2 and 3. The product of each of the single digits is written in the corresponding square as shown below. Once numbers in the lattice have been written, the final product, 1104, is formed by adding along the diagonals from upper right to lower left, being careful to remember to carry.

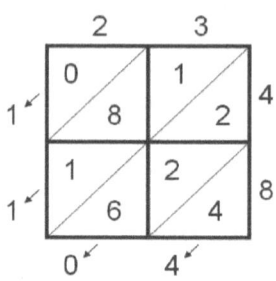

Another method still taught in the early twenty-first century has been called "cross-multiplication." It was described by Leonardo Fibonacci in the twelfth century but was certainly known earlier in India and the Middle East. To multiply 23×48, one starts from the right and multiplies 3×8 to get 24. The 4 is written down and the 2 is remembered. Then the cross multiplication is performed, $2 \times 8 + 3 \times 4$ to get 28, which is added to the remembered 2, obtaining 30. The 0 is written down and the 3 is remembered. Finally 2×4 is computed obtaining 8, which is added to the remembered 3, resulting in 11, which is written down. The final result is thus 1104. This method can be generalized and, by keeping various "remembered" digits using finger numbers, it is possible to multiply many two-digit numbers without writing down any intermediate results.

Probably the most popular method for multiplying that is taught in the early twenty-first century computes the products of multiplicand with each of the digits of the multiplier, working from right to left. Each of these successive products is shifted one more digit to the left. These partial products are then summed to obtain the final result as the following example shows:

```
      4 8
    x 2 3
    1 4 4
  + 9 6
  1 1 0 4
```

This and similar methods were implemented on various counting boards, the abacus, and dust boards.

Division Algorithms

An ancient method for finding the quotient of two large whole numbers that is most appropriate for either the dust board or the Chinese counting board was adapted to pen and paper and became the "scratch method" that was used in Europe up into the nineteenth century.

Two methods of computing long division using pencil and paper eventually replaced the scratch method.

One method is popular in Italy, England, and the United States and will be referred to as the "Italian method." The other is popular in Spain, France, Latin America, Austria, and Germany. This second approach will be referred to as the "Spanish method." Both methods date from at least the dawn of the sixteenth century. Both methods are shown by demonstrating how to find 2456/57, which is 43 with a remainder of 5.

```
        Italian Method              Spanish Method

              4  3                 2 4₃5  6 | 5 7
       ┌─────────                  1 7₂6  4   4 3
    5 7│2  4  5  6                       5
      -2  2  8
      ─────────
           1  7  6
          -1  7  1
          ────────
                 5
```

In the Italian method, the dividend is written under a horizontal line above which the quotient will be written as it is found. The divisor is written to the left of the dividend with a vertical line drawn between them. It is determined that 57 will go into 245 at least (but no more than) 4 times, and 4—the first digit of the quotient—is written above the 5 of the dividend. The product $4 \times 57 = 248$ is computed and written below the 245. Subtraction is performed, obtaining 17 and the 6 from the dividend is brought down to the right of the 17. Then, it is determined that 57 will go into 176 at least (but no more than) 3 times. The product $3 \times 57 = 171$ is computed and subtracted from 176 to get the remainder of 5.

The Spanish method is cosmetically different and is characterized by many fewer digits being written down because the multiples $4 \times 57 = 248$ and $3 \times 57 = 171$ are never explicitly computed. To start, the dividend is written down followed by a long vertical line and the divisor. A horizontal line is then drawn under the dividend and divisor. The quotient will be developed to the right of the vertical line below the horizontal line one digit at a time. First, it is determined that the first digit of the quotient is 4. Then $4 \times 7 = 28$ is computed, which must be subtracted from 5. This operation cannot be done, and so a little 3 is written between the 4 and the 5 of the dividend. Now 28 can be subtracted from 35 obtaining 7, which is written under the 5 of the dividend. The 4×5 is found and added to the little 3 obtaining 23. This number is subtracted from 24 obtaining 1, which is written below the 4 of the dividend. To complete this phase, the 6 from the dividend is brought down so that the problem is now to divide 176 by 57. It is determined that 57 will go into 176 at least (but no more than) 3 times, and so 3 is written down as the next digit of the quotient. Computing $3 \times 7 = 21$, try to subtract 21 from 6, which cannot be done, and so a small 2 is written between the 7 and 6. Now, subtract 21 from 26, obtaining 5, which is written below the 6. Now, 3×5 is computed and added to the little 2, obtaining 17, which when subtracted from 17 is 0. As such, nothing needs to be written to the left of the 5. By the time a student is out of elementary school, it is expected that the annotations like the little 2 and 3 will no longer need to be written.

Checking Results

Because the algorithms for computing products and quotients of whole numbers are complex, methods to check the results have existed since antiquity. Since division and multiplication can be viewed as the inverse operations of each other, a division can be checked by multiplying the quotient by the divisor and adding the remainder. If the result is the dividend, then the division has been performed correctly. Another approach to checking the result of an arithmetic operation is to perform the operation using modular arithmetic, typically with respect to 7, 9, or 11. A number a modulo b is defined to be the remainder of $a \div b$. Thus, 267 modulo 9 is 6. There are easy tricks for computing numbers modulo 7, 9, and 11. For example, the method of "casting out 9s" can be used to compute 267 modulo 9 by simply summing the digits and subtracting 9 whenever the sum is over 9. For example, $2 + 6 + 7 = 15 - 9 = 6$. In order to check the correctness of the multiplication $23 \times 48 = 1104$, check to see if $[(23 \text{ modulo } 9) \times (48 \text{ modulo } 9)]$ modulo $9 = 1104$ modulo 9.

In this case, $5 \times 3 = 15$ and 15 modulo 9 is 6, which is equal 1104 modulo 9. One can conclude that the product, 1104, is probably correct. Although it is possible that such a check will confirm an incorrectly performed multiplication, this is unlikely.

Multiplication by Addition

Because multiplication and division are time consuming and tedious to perform compared to addition and subtraction, there has been much work to simplify the finding of products and quotients. Simplification results from using logarithms, invented in the seventeenth century, since

$$\log(A \times B) = \log(A) + \log(B).$$

Thus, it is easy to find the product of two numbers using a table of logarithms by looking up the logarithm of both the multiplier and the multiplicand in the table, adding these two logarithms and then using the table to find the number that has that sum as its logarithm.

It is also possible to convert multiplication to addition (and halving) using a table of cosines along with the trigonometric identity

$$\cos(A)\cos(B) = \frac{\cos(A+B) + \cos(A-B)}{2}.$$

This method was used by some astronomers before the invention of logarithms.

Rapid Multiplication and Division

Throughout history, people have developed the ability to multiply large digits in their head. Thomas Fuller, a slave, was one such person. For example, he calculated the number of seconds a man who is 70 years, 17 days, and 12 hours old has lived. He correctly answered 2,210,500,800 in only a minute and a half, and historians hypothesize that the algorithms he used were probably based on traditional African counting systems. In the twenty-first century, mathematician and magician Art Benjamin has turned his rapid mental calculations into educational entertainment.

Multiplication of whole numbers that can be represented as single binary words in a computer can typically be done with a single hardware instruction that combines addition and shifting to find the product. To multiply numbers with thousands of digits, other methods are possible that make use of fast methods for computing Fourier transforms, named for Joseph Fourier. The Schönhage–Strassen algorithm, developed by Arnold Schönhage and Volker Strassen, and the Fürer algorithm, developed by Martin Fürer, are two such methods. These complicated algorithms, however, have limited practicality when implemented on a conventional computer. Mathematicians in the field of numerical analysis consider issues of speed and error in computer algorithms.

Early computers multiplied or divided using repeated addition, subtraction, and shifting algorithms. Depending on the computing system, the amount of time required for multiplication or division on a computer is approximately the same. However, factoring a number into its unknown divisors is significantly harder. The RSA (which stands for the people who first described the system: R. Rivest, A. Shamir, and L. Adleman) cryptosystem takes advantage of this characteristic.

Generalizing Multiplication and Division

Multiplication and division can be used for computational purposes and to help understand other mathematical principles. The product of two rational numbers a/b and c/d is defined to be the rational number

$$\frac{ac}{bd}$$

whereas the quotient, $a/b \div c/d$ is defined be the product of a/b and d/c. The reciprocal of a rational number, a/b, is the number b/a, since the product of a/b and a/b is 1. The product of two irrational numbers, which cannot be represented as a fraction of whole numbers, is approximated by taking the product of their approximating rational numbers. The product of irrational measurements can be found exactly with geometry using similar triangles.

Multiplication can be generalized to other mathematical objects, such as complex numbers and matrices. One of these objects, typically called the "identity" and denoted by "1," has the property such that if a is any object, then the product of 1 and a is a. The reciprocal of an object a, if it exists, is denoted by a^{-1} and is defined to the object so that the product, $a \times a^{-1}$ is 1. When the reciprocal of an object b exists, then the quotient of objects a and b is defined to be $a \times b^{-1}$.

Further Reading

Aho, A. V., J. E. Hopcrotf, and J. D. Ullman. *The Design and Analysis of Computer Algorithms.* Reading, MA: Addison-Wesley, 1976.

Boyer, C. B. *A History of Mathematics*. Hoboken, NJ: John Wiley and Sons, 1991.

Gillings, R. J. *Mathematics in the Time of the Pharaohs*. New York: Dover Publications, 1982.

Smith, D. E. *History of Mathmatics, Volume II*. New York: Dover Publications, 1958.

CARL R. SEAQUIST
CATHERINE C. GALLEY

Normal Distribution

Category: History and Development of Curricular Concepts.
Fields of Study: Calculus; Communication; Connections; Data Analysis and Probability.
Summary: Better known to laymen as the bell curve, there are many applications for normal distribution.

The normal distribution is one of the most useful and important probability distributions, with a wide range of theoretical and real-world applications. Many people know the normal distribution primarily by its colloquial name, the "bell curve," which comes from its characteristic shape: a symmetric curve with a pronounced peak in the middle and diminishing tails. Mathematically, normal distributions are a family of continuous probability distributions. The normal function has no closed-form integral, but areas under the curve, which correspond to probabilities, can be accurately approximated with methods like numerical integration. All normal probability distributions display the same symmetric bell shape, but can have any real-valued mean (μ) and positive real-valued standard deviation (σ). The standard normal distribution is a special case with a mean of zero and standard deviation of one. All normal distributions can be transformed or standardized to the standard normal, which is theoretically important and extensively tabulated. Computers and calculators also allow direct calculation of normal probabilities. Students often use both technology and tables when they study the normal distribution in high school and beyond.

Many naturally occurring phenomena are normally or approximately normally distributed, like the heights of adult human beings. In other cases, such as intelligence tests, the measurements are purposely structured or scaled according to this distribution. Several other probability distributions converge to the normal distribution or are well approximated by it. The central limit theorem, based on normal approximations, is the foundation for a wide range of commonly used statistical procedures, particularly for estimation and inference. Another common name for the normal distribution is the "Gaussian distribution," after Carl Friedrich Gauss, whose work significantly advanced many statistical theories and concepts. Occasionally it is referred to as the "Laplace distribution," after Pierre-Simon Laplace. The variety of names for the normal distribution likely reflects the debate on the origins of the term "normal distribution" and the breadth of people who influenced its development.

History

The first appearance of the term "normal distribution" in a published document is often credited to a seminal paper from Karl Pearson in 1895. However, there are some who say the first use corresponds to Charles

Galileo and Bernoulli

The early origins of the normal distribution can be traced in part to Galileo Galilei and his work in astronomy. In 1610, Galileo noticed that the measurement errors in astronomical tables were distributed symmetrically (in an unbiased fashion) around the correct value.

A century later, Jacob Bernoulli made two critical advances toward the development and characterization of the normal distribution. The first was the "law of large numbers" (named as such by Simeon Poisson in 1835). The second was the development of the binomial distribution. The law of large numbers predicts the convergence of sample means to the true population mean as sample size approaches infinity. The binomial distribution models probability in situations in which there are sequences of independent random events with two equally likely outcomes for each event, such as flipping a coin.

Peirce in 1783, to Francis Galton in 1889, or to Henri Poincaré in 1893. Statistician and historian Stephen Stigler believes that it might have been used much earlier, and there is certainly evidence to support that assertion.

Abraham DeMoivre is credited with the first mathematical derivation of the normal distribution in his 1733 work *Approximatio ad summam terminorum binomii (a+b)^n in seriem expansi*. Using sums of Bernoulli's binomial random variables, he approximated a continuous distribution to the discrete binomial using integral calculus, which resulted in a bell-shaped continuous distribution. Continuing this idea, Pierre-Simon Laplace presented the central limit theorem in 1778, which is also sometimes called the "DeMoivre–Laplace theorem." In fact, the name "central limit theorem" is credited to George Pólya's 1920 work on the normal distribution. Since the central limit theorum is the limit of a summation of binary variables, it is applicable to both discrete and continuous random variables. It has many real world applications along with its theoretical importance, and it is fundamental to statistical inference.

Robert Adrain, an American, and Carl Friedrich Gauss, a German, worked simultaneously on similar notions at the start of the nineteenth century without being aware of each other's work. In 1808, Adrain presented arguments regarding the validity of the normal distribution for describing distributions of measurement errors, inspired by a real-world problem in surveying. He used this initial work to further develop and prove Adrien-Marie Legendre's method of least squares. Gauss published his *Theory of Celestial Movement* in 1809. This work included several critical contributions to mathematics and statistics, including the maximum likelihood parameter estimation, the method of least squares, and the normal distribution. This is perhaps part of the reason that Gauss tends to be given credit over Adrain for their similar contributions regarding the normal distribution.

In 1829, Adolphe Quetelet brought the concept of the normal distribution of error terms into the analysis of social data. He wanted to discover the underlying laws of society in the same way other researchers were exploring scientific and mathematical laws. Quetelet invented the term "social physics" and empirically developed the first notions of the measure now called "body mass index." He analyzed several data sets of human biological and social data, such as the heights and weights of conscripted soldiers, and by inductively using the central limit theorem, he concluded that the normal error distribution described these measures quite well. Galton also contributed to the application and development of the normal distribution in the biological and social sciences. He produced the first known index of correlation as well as regression analysis, and he proved that a normal mixture of normal distributions is itself normal. His colleagues Walter Weldon and Karl Pearson also contributed to normal theory and applications, and the three of them cofounded the journal *Biometrika*. The field of biometrics is generally traced back to Weldon's seminal papers. Pearson used the method of moments to estimate mixtures of normal distributions and further developed correlation and regression methods based on the normal distribution. However, part of his motivation for developing methods like chi-square analyses was apparently to try to decrease the growing reliance on the normal distribution as a foundation of statistical theory and analytic methods.

Pearson's efforts to diminish the role of the normal distribution in statistics failed. Many other mathematicians and statisticians, including Pearson's son Egon, continued to develop theory and applications in a variety of areas. For example, William Gossett and Ronald Fisher derived and refined the closely related Student's *t* distribution in the early twentieth century. The distribution is not called Gosset's *t* because he worked for Guinness Brewery and he could not publish his work in his own name because of proprietary issues, so he adopted the pseudonym "Student." Starting in the 1930s, Samuel Wilks explored many aspects of normal distributions. These included deriving sampling distributions for parameter estimates in bivariate normal distributions as well as for covariances in multivariate normal distributions, which led to important advances in multivariate statistical methods. The American Statistical Association's Wilks Award is one of the most prestigious in the field of statistics. Miroslaw Romanowski published a generalized theory of modified normal distributions in 1968 that help characterize errors that do not seem to be well-described by the normal distribution. Another such generalization is the skew normal. Other related distributions include the "lognormal distribution" or "Galton distribution," which describes a variable whose log is normally distributed, and the "folded normal," which is based on taking the absolute value of a normal distribution.

Recent Developments

The term "bell curve" became even more widely known in 1994 when psychologist Richard Herrnstein and political scientist Charles Murray wrote *The Bell Curve*, which took its name from the distribution of IQ scores and included a picture of the normal distribution on its front cover. Herrnstein and Murray correlated intelligence scores with social outcomes and asserted that social stratification based on intelligence was on the rise. The book remains highly controversial for the authors' inclusion of discussions regarding supposed relationships between race and intelligence and has spurred many debates on both social and statistical matters.

Further Reading

Heyde, Chris, and Eugene Seneta, eds. *Statisticians of the Centuries*. New York: Springer-Verlag, 2001.

Stigler, Stephen M. *The History of Statistics*. Cambridge, MA: Harvard University Press, 1986.

Carlos J. Vilalta

Number and Operations

Category: History and Development of Curricular Concepts.
Fields of Study: Communication; Connections; Number and Operations.
Summary: Numerous civilizations throughout history developed unique number systems and number operation methods, some principles of which survive into the twenty-first century.

The properties of numbers and operations are among the first concepts that most people learn about mathematics and they were also among the earliest type of mathematical knowledge developed historically. Number and operations are pervasive in school curricula. There are many types of numbers (for example, integers, irrational numbers, and imaginary numbers), each with their own properties. Learning how to work with different types of numbers is basic to the work of learning mathematics. The term "operations" refers to the practice of applying some rule on a set of numbers; the four basic operations are addition, subtraction, multiplication, and division. However, there are a wide variety of other mathematical operations or operation-like procedures on many types of mathematical objects, such as modular arithmetic, that may be explored at many levels.

In the twenty-first century, students in the earliest grades start to investigate whole numbers and common fractions along with addition and subtraction. In the later primary grades, they may study base-10 decimals, a broader range of fractions, negative numbers, and equivalent forms for fractions, decimals, and percentages. Operations extend to include addition and subtraction of common fractions or decimals; multiplication and division of whole numbers; and relationships between operations. Concepts like ratios and proportions, integers, factorization, prime numbers, and some alternative methods of notation for very large numbers begin to be introduced in middle school. Students learn more arithmetic procedures with fractions, decimals, or integers, as well as to simplify computations using addition and multiplication properties. They also investigate squares and square roots. In high school, students may study very large and small numbers; properties of numbers and various number systems; vectors and matrices with real number properties; and number theory. Operations begin to include addition and multiplication of vectors and matrices, as well as permutations and combinations. These concepts continue to be extended into college with new systems of numbers and operations or operation-like procedures.

Early Number Systems

The first type of numbers people generally learn about are called the "natural" or "counting numbers": 1, 2, 3, 4, 5, 6 . . . The historical record shows that the use of counting numbers is an ancient practice. As with measurement, body parts may have been used, and archaeologists have found other evidence, such as notched bones, that support the idea of tallying or counting. The Egyptians used a base-10 system, written either with hieroglyphs or hieratic (cursive) script, and using special symbols for powers of 10 (10, 100, 1000, and so on). The ancient Egyptians were also aware of fractions, which were primarily written as unit fractions of the form 1/*n*, such as 1/2 or 1/4, although some Egyptian texts contain fractions of the form 2/*n* or 3/*n*, and other quantities were expressed as combinations of unit fractions.

The Babylonians used a numerical system with a base of 60, a practice that survives into the twenty-first century in the convention of dividing a circle into 360 degrees and in units of time, such as 60 seconds in a minute and 60 minutes in an hour. They used only two symbols, one signifying 1 and the other 10, to write all the values 1–60, and used the place system so that the meaning of a symbol depended on its place within a number—a major advance that was crucial to the development of modern mathematics. In a decimal or base-10 system,

$$111 = \left(1 \times 10^2\right) + \left(1 \times 10^1\right) + \left(1 \times 10^0\right).$$

The same digits in the base-60 system would mean

$$111 = \left(1 \times 60^2\right) + \left(1 \times 60^1\right) + \left(1 \times 60^0\right)$$

or the same quantity as 3661 in a base-10 system.

The ancient Mayan developed a number system with a base of 20 and a place system, using dots (with a value of 1) and bars (with a value of 5) to write the numbers 1–19, with powers of 20 indicated vertically. The Mayans also understood the concept of zero as a placeholder and had a special symbol for it, which they used in their calendar system.

An acrophonic number system was used in Greece by the first millennium B.C.E. Acrophonic means that numbers are signified by the first letter of the word used for that number, with symbols for 1, 5, 10, 100, and so on. As with the more familiar Roman numerals, this system was an additive system (rather than place), so the value of a number was found by adding up the value of all the symbols that comprised it. A competing system also used in Greece was one in which each letter of the alphabet was assigned a numeric value reflecting its order in the alphabet. In this system, the first 10 letters (*alpha* through *iota*) correspond to the numbers 1–10, then the next letter (*kappa*) stands for 20, the next (*lambda*) for 30, until *rho*, which signifies 100. The next letter (*sigma*) signifies 200, and so on. This system was also an additive system, so that 12 was written as *iota beta* or 10 + 2 and 211 as *epsilon iota alpha*. Numbers 1000–9000 were written by adding a superscript or subscription to the letters *alpha* through *theta*, while larger numbers were written with the symbol *M* (meaning "myriad") for 10,000, with multiples indicated by writing other numbers above the *M*.

Roman Numerals

The familiar system of Roman numerals was developed from about the third century B.C.E. It was used throughout the Roman Empire and in Europe into the Middle Ages and was eventually replaced by the more efficient Hindu–Arabic number system. The Roman number system has the benefit of using only a few symbols, but does not include the concepts of zero or of place, so the value of a number is calculated by adding together all the values of its elements. The symbols used include M for 1000, D for 500, C for 100, L for 50, X for 10, V for 5, and I for 1, with the later refinement that a smaller number could be placed next to a larger number to indicate subtraction. Roman numerals translate to Hindu–Arabic numerals as the following:

LXXIII = 73
CDXXXII = 432
MCMLXXXV = 1985
MMX = 2010

Roman numerals are still in use in the twenty-first century to indicate succession (for example, King Richard III of England) and sometimes in film release dates. The inefficiency of the Roman system compared to the modern system of Hindu–Arabic numerals can be illustrated by trying to quickly determine which of the following three dates is most recent: MCMXCIX, MCMLXXXVII, and MMVII. Now try again with the same values in Hindu-Arabic numerals: 1999, 1987, and 2007.

Indian or Hindu Numerals

Indian or Hindu numerals and the concept of zero (written as a dot or small circle and referred to by the Sanskrit term *sunya*, which means "empty") also appear to date to the third century B.C.E. Historians have cited Brahmi numerals, which share a name with a family of alphabets or scripts; which evolved into Gupta numerals, named for the fourth to sixth century C.E. Gupta dynasty; and then Nagari or Devanagari numerals, also named for alphabet systems, beginning in about the ninth century; and finally symbols that looked very much like the familiar numerals 0–9 somewhere around the fourteenth century. There are many origin theories for Hindu numerals, which fall into two general classes: they came from an alphabet (as did the Greek system) or they came from some other earlier number system (as did Roman numerals). Hindu number systems were predominantly base 10, and documents suggest that Indians were using a place value system by the sixth century C.E. Mathematician Pierre-Simon Laplace said, "The ingenious method of expressing every possible number using a set of ten symbols (each symbol having a place value and an absolute value) emerged in India . . . Its simplicity lies in the way it facilitated calculation and placed arithmetic foremost amongst useful inventions."

Hindu systems of numerals appear to have made their way into Arabic and Islamic cultures in the latter half of the first millennium C.E. These Hindu numerals, along with a base-60 system using Arabic letters to represent numbers (common among astronomers) and a "finger arithmetic" system (widely used in business), coexisted for some time in the Arabic world. In the ninth century, mathematician Abu Ja'far Muhammad ibn Musa Al-Khwarizmi wrote *On the Calculation with Hindu Numerals*. His contemporary Abu Yusuf Ya'qub ibn Ishaq al-Kindi also wrote *On the Use of the Indian Numerals* (c. 830 C.E.). Several Arabic and Islamic scholars studied Hindu numerals in the tenth century. Abu Ali al-Husain ibn Abdallah ibn Sina (also known as Avicenna) was purportedly taught by Egyptians, and Abu'l Hasan Ahmad ibn Ibrahim Al-Uqlidisi is credited with helping to modify Hindu numerals to replace the traditional "finger arithmetic." Mathematician Abu Arrayhan Muhammad ibn Ahmad al-Biruni visited India in the eleventh century C.E., though even before his first travels he had examined Arabic translations of Indian mathematics texts.

Hindu–Arabic Number System

The Hindu–Arabic number system was adopted in Europe a few centuries later, replacing Roman numerals, as Europeans became familiar with Arabic manuscripts. The first known example of Hindu numerals in a European document are in the tenth-century *Codex Vigilanus*, but the beginnings of widespread use appear to date closer to the fifteenth century. The symbols used in this system are similar to those used in Europe in the twenty-first-century (0–9), while a different set of symbols is used with the same number system in the Middle East and in parts of India (thus many Arabic speakers do not use what in the United States are commonly called "Arabic numerals"). The Moroccan mathematician Abu Bakr Al-Hassar is credited with developing the modern method of notating fractions (two numbers separated by a horizontal bar) in the twelfth century. *Liber Abaci*, written by Italian mathematician Leonardo Fibonacci in the early thirteenth century, was also influential in spreading the use of Hindu–Arabic numerals (and the place system) throughout Europe.

Any number may be used as a base when numbers are written using the place system. For instance, the binary (base 2) and hexadecimal (base 16) systems are used in work with computers. In the binary system, there are two digits (0 and 1), and each successive place is a greater power of 2. In the hexadecimal system, there are 16 digits (letters are used to express the extra digits required, so A = 10, B = 11, C = 12, D = 13, E = 14, and F = 15). If necessary to avoid confusion, the base of the number system may be included as a subscript; for example, 17_{10} would be 17 in the base-10 system.

Modular arithmetic was developed by German mathematician Leonhard Euler in the eighteenth century and was advanced by others, including German astronomer and mathematician Carl Friedrich Gauss, whose 1801 *Disquisitiones Arithmeticae* established many of the rules of number theory. Modular arithmetic is sometimes called "clock arithmetic" because the concept is similar to that of a 12-hour clock. If it is currently 9 o'clock, 6 hours later it will be 3 o'clock, not 15 o'clock, because the clock starts over with 0 as soon as it reaches 12.

Aids to Computation

Systems to aid computation are almost as old as number systems themselves. For instance, Egyptian scribes used tables to help them perform arithmetic with fractions, and the abacus or counting frame was used in several ancient cultures, including those of Mesopotamia, Egypt, Persia, and Rome. However, the abacus is most strongly identified today with Asia, in particular China, where it was used at least as early as the second century B.C.E. A Chinese abacus consists of a number of rods divided by a beam into two regions or decks: the upper deck of each rod has two beads, and the lower deck has five. Mathematical operations are carried out by sliding the beads toward or away from the deck, and expert abacus operators can rapidly solve problems involving not only the four basic functions (addition, subtraction, multiplication, and division) but also square and cube roots. The simplicity and efficiency of the abacus encouraged its spread to other Asian countries, including India, Japan, and Korea. Use of the modern Japanese abacus, which uses one bead in the upper deck and four in the lower, is still taught in primary schools in Japan today as it is believed to aid students in forming a mental representation of numbers.

Arabic mathematicians developed a system of lattice multiplication, which involves using a lattice or grid of boxes divided into diagonal halves. To perform lattice multiplication, the two numbers to be multiplied are written across the top and the side of the grid, the digits are multiplied separately and then added along the diagonals to produce the result. This system was introduced to Europe by Fibonacci in 1202. It was improved by Scottish mathematician John Napier in the early seventeenth century through a type of abacus referred to as "Napier's bones," which consists of a tray and a set of 10 rods, one for each digit 0–9. Each rod is divided into nine squares, with each but the top divided by a diagonal line. Each square contains the product of its own digit multiplied by each other digit; for instance, the rod for 5 contains the values 5, 1/0, 1/5, 2/0, and so on (the / indicating the diagonal of the square). Napier's bones are used to multiply, divide, and extract square roots. For example, to multiply, the rods for one number are placed in the tray, and the values from the rows comprising the digits of the second number are read off, adding together the pairs of values on the diagonals.

Logarithms are another important aid to calculation. A logarithm is an exponent such that when the base of a number system is raised to that power, the result will be the number. For instance in base 10, the logarithm of 100 is 2 because $10^2=100$. In the system of natural logs, the base is e (sometimes called "Euler's number" after the Swiss eighteenth-century mathematician Leonhard Euler), the irrational constant 2.718281.... The natural log of 100 is 4.6051 because $e^{4.6051} \approx 100$.

One common use of logarithms before the advent of electronic calculators and computers was to simplify multiplication, division, and the calculation of powers and roots. As such, logarithms played an important role in the development of astronomy and other mathematically based sciences.

Napier is usually credited as the inventor of the logarithm due to his 1614 publication *Mirifici Logarithmorum Canonis Descriptio*, which included tables of natural logarithms and explanations of their use. Important tables of base 10 logarithms were published in 1617 and 1624 by English mathematician Henry Briggs.

Multiplication using logarithms rests on the following rule. For any base b

$$c \times d = b^{(\log_b c + \log_b d)}.$$

For instance, if the base is 10, c is 108, and d is 379:

$$108 \times 379 \approx 10^{(2.033424 + 2.578639)} \approx 10^{4.612063} \approx 40,932$$

because $108 \approx 10^{2.033424}$ and $379 \approx 10^{2.578639}$.

Conducting multiplication in this way requires only looking up the two logarithms in the table, adding them, and looking up the antilogarithm (the base 10 raised to a power) in another table, which for large numbers is much quicker than doing the multiplication by hand. The slide rule, also developed in the seventeenth century, made the process even quicker and remained in common use well into the twentieth century.

Further Reading

Dantzig, Tobias. *Number, the Language of Science: A Critical Survey for the Cultured Non-Mathematician*. New York: Doubleday, 1956.

Eymard, Pierre, and Jean-Pierre Lafon. *The Number Pi*. Translated by Stephen S. Wilson. Providence, RI: American Mathematical Society, 2004.

Flannery, David. *The Square Root of 2: A Dialogue Concerning a Number and a Sequence*. New York: Copernicus, 2006.

Hodger, Andrew. *One to Nine: The Inner Life of Numbers*. New York: W. W. Norton, 2008.

Kaplan, Robert. *The Nothing That Is: A Natural History of Zero*. New York: Oxford University Press, 2000.

Sarah Boslaugh

Number Theory

Category: History and Development of Curricular Concepts.
Fields of Study: Communication; Connections; Number and Operations.
Summary: Number theory has captured the imagination through numerous famous problems, many still unsolved.

The legendary mathematician Carl Friedrich Gauss (1777–1855) famously described number theory as "the queen of mathematics." The core of number theory is the study of the integers, but number theory includes a much wider class of concepts and problems that arise in the study of the integers. Number theory is extremely popular as recreational mathematics and is explored in twenty-first-century high school classrooms. Early mathematicians in Greece, India, and the Islamic world investigated and developed number theory, making enormous and widely varied contributions. The field has continued to blossom through the twenty-first century. Some mathematicians view number theory as a branch of pure mathematics, but others view it as applied mathematics because of its utility to so many fields, such as physics, chemistry, biology, computing, engineering, coding, cryptography, random number generation, acoustics, communications, graphic design, music, and business.

Prime Numbers

A large portion of number theory is related directly or indirectly to the study of prime numbers and divisibility. An integer a is divisible by integer b if there is an integer c such that $a = bc$. A prime number is an integer $p > 1$ that is divisible only by itself and 1, and a composite number is a positive integer with more than two factors. It is technically most convenient to consider 1 to be neither prime nor composite. The so-called Fundamental Theorem of Arithmetic, investigated by Carl Friedrich Gauss in the nineteenth century, states that every positive integer $n > 1$ can be written as a product of prime numbers, and furthermore, that this prime factorization is unique (except for the order in which the factors are written). The theorem was partially proved by Euclid of Alexandria in ancient Greece. The recognition that the theorem does not hold in more general number systems by mathematicians such as Ernst Kummer led to the development of the field of algebraic number theory.

It is well known that there are infinitely many prime numbers, so it would not be possible to obtain a complete list of all prime numbers. The search for ever-larger prime numbers is ongoing, and testing numbers for primality is sometimes used as a test of the computational power of supercomputers.

Modular Arithmetic and Cryptography

An important component of elementary number theory is modular arithmetic. In modular arithmetic, two numbers are treated to be the same if they have the same remainder when divided by some given number, the "modulus." One writes $a = b \pmod{m}$ if m divides the difference $a - b$. One reason why this concept is so useful is that it is compatible with the operations of addition, subtraction, and multiplication. If $a = b \pmod{m}$ and $c = d \pmod{m}$, then $a + c = b + d \pmod{m}$, $a - c = b - d \pmod{m}$, and $ac = bd \pmod{m}$. The situation is complicated for division, unless m is a prime. Modular arithmetic is sometimes called "clock arithmetic" because of the similarity between arithmetic mod 12 and the system for counting hours. One related theorem that is frequently studied in classrooms in the Chinese remainder theorem, named so because the theorem originates in Chinese texts by Sun Tzu and Qin Jiushao.

If there was ever a time when number theory was studied *only* for its elegance and beauty and not for any application, that time is past now. Modern cryptography, essential for the security of the Internet, is based heavily on number theory. The widely used RSA (Rivest, Shamir, and Adleman) encryption methods are based on modular arithmetic and rely for their security on number theorists' understanding of how difficult it is to find large prime factors of a large number. Other encryption systems based on more exotic number theoretic objects, such as elliptic curves, are under active development by cryptographers and number theorists.

Algebraic and Analytic Number Theory

Number theorists study in a diverse range of fields related to number theory, including probabilistic number theory, diophantine approximations, the geometry of numbers, and combinatorial number theory. Number theorists do not exclusively study the integers, nor are the integers the only system that admits something like prime numbers. Number theorists also extensively study other systems of algebraic numbers. An algebraic number is a number that is the root of a polynomial with integer coefficients; the role of "integer" is played by numbers that are roots of polynomials with integer coefficients and leading coefficient 1. For example, the square root of 10 is an integer in this extended sense. Much of algebraic number theory concerns how the concepts of prime number and divisibility as applied to other number fields compare and contrast with the familiar situation for the standard integers.

For example, the Gaussian integers are those complex numbers $a + bi$ with a and b both integers. The Gaussian integers form a number system in which the concepts of prime number and divisibility above apply almost exactly as described above. However, the set of prime numbers here is very different. The number 7, for example, is still prime in the Gaussian integers, but 13, which is prime in the integers, factors here as the product of the two primes, $3 + 2i$ and $3 - 2i$.

Though the arithmetic integer is apparently part of discrete mathematics, there is a large branch of number theory, called "analytic number theory," which applies extremely sophisticated techniques from calculus and complex analysis to the problems of number theory.

A basic fact of analytic number theory is that the sum of the reciprocals of all primes,

$$\sum \frac{1}{p}, \text{ is infinite.}$$

Note that it can be recovered from this that the set of prime numbers must itself be infinite. With care, one can estimate the sum of the reciprocals of all primes up to some number x, which turns out to grow at the same rate as $\log(x)$. With more refinement along these lines, one obtains the celebrated Prime Number Theorem, which says that the number of prime numbers less than some large integer x is well-approximated by

$$\frac{x}{\log(x)}.$$

Famous Problems in Number Theory

One feature of number theory is a large number of intriguing problems that are very simply stated but require unexpectedly advanced and specialized techniques to solve; many remain unsolved in the early twenty-first century. Indeed, many or most of the major problems in mathematics that are known to the general public have origins in number theory.

One major recent mathematical breakthrough, which received mainstream media coverage, was the proof of Fermat's Last Theorem. Pierre de Fermat (c. 1601–1665) wrote a note in the margin of a book he was reading to the effect that there are no integral solutions to the equation $x^n + y^n = z^n$ with $n > 2$. There are infinitely many solutions with $n = 2$, and these have been well studied; by the Pythagorean Theorem they correspond to right triangles with integer-length sides. Fermat never wrote down a proof, writing instead that the margin of his book was too small to contain it. In the intervening centuries, mathematicians tried to supply the missing proof. Much of algebraic and analytic number theory was developed as part of the effort to prove this theorem. The problem was finally solved in 1995 by Sir Andrew Wiles (1953–), using extremely sophisticated number-theoretic objects involving elliptic curves and modular form. It is not now generally believed that Fermat ever possessed a valid proof.

Because all primes other than 2 are odd, the smallest possible difference between consecutive primes (other than 2 and 3) is 2. Prime numbers that differ by 2 (those that are as close as possible) are called "twin primes." The Twin Prime Conjecture asserts that there should be an infinite number of twin primes, but mathematicians are very far away from being able to prove this. In a triumph of analytic number theory, Viggo Brun (1885–1978) showed that the sum of the reciprocals of all twin primes is finite, though mathematicians still do not know whether there are finitely or infinitely many of them!

Another famous problem about the additive distribution of the set of prime numbers is Goldbach's Conjecture, named for Prussian mathematician Christian Goldbach (1690–1764). The conjecture asserts that every even number larger than 2 can be written as the sum of two (not necessarily different) prime numbers. Computer searches have verified that there are no counterexamples smaller than one quintillion. This problem has occupied the attention of many recreational mathematicians and has been featured in several novels and television shows. Again, mathematicians are very far from proving such a theorem.

The Riemann Hypothesis is the most important open problem in number theory, and arguably in all

Numbers, Complex

Category: History and Development of Curricular Concepts.
Fields of Study: Algebra; Communication; Connections; Number and Operations.
Summary: Complex numbers inevitably arise in many situations, but may be difficult to accept.

Complex numbers are ubiquitous in modern science, yet it took mathematicians a long time to accept their existence. They are numbers of the form $z = a + bi$ where a and b are real numbers, and i is a symbol called the "imaginary unit," which satisfies the seemingly impossible equation $i^2 = -1$. The numbers a and b are called the "real" and "imaginary parts" of z, respectively. The imaginary unit can be thought of as the square root of -1 and is also written $i = \sqrt{-1}$. In fact, any negative number has a complex square root; for example, the square root of -15 is the complex number

$$\sqrt{-15} = \sqrt{15} \cdot i.$$

In the twenty-first century, science students routinely encounter complex numbers, for instance, as solutions to quadratic equations.

In mathematics, complex numbers form an independent area of research and are also used to prove theorems in other areas of mathematics; examples are Machin's formula and the Prime Number theorem. In the natural sciences, complex numbers often simplify calculations, for example, in the theory of relativity where the distance between points in space-time can be computed using imaginary time coordinates. The complex exponential function is used in electrical engineering as a convenient way of simultaneously describing the amplitude and phase of an alternating current, and in chaos theory, the complex plane is the scene of computer-generated fractals, such as the "Mandelbrot set," named after Benoît Mandelbrot.

Unlike natural numbers, which are used for counting, and real numbers, which are used for measuring distances, complex numbers have no obvious real-life interpretation. For this reason, the questions of what complex numbers really are remained a controversial topic for three centuries after their discovery in the sixteenth century. The term "imaginary numbers"

of mathematics. Named for Bernhard Riemann (1826–1866), this conjecture concerns a particular function of the complex numbers called the *zeta function*. Let $\zeta(s)$ be the value of the infinite sum

$$1^s + \frac{1}{2^s} + \frac{1}{3^s} + \frac{1}{4^s} + \frac{1}{5^s} + \cdots.$$

This is apparently defined only for real numbers $s > 1$, but it turns out that there is a uniquely meaningful way to extend this to allow any complex number as input. It is relatively easy to show that

$$\zeta(-2) = \zeta(-4) = \zeta(-6) = \cdots = 0$$

and there are infinitely many other "nontrivial zeroes" s such that $\zeta(s) = 0$. The standing conjecture is that all the nontrivial zeroes of ζ lie on a particular line in the complex plane. Though a tremendous amount of effort has gone into trying to prove this and though mathematicians have much corroborating evidence, it is still open. The statement might seem esoteric and arcane; surprising as it may seem, this statement, if true, would have profound implications about the distribution of prime numbers, which would have ramifications throughout all mathematics. Mathematicians have found dozens of very different-looking statements that are known to be equivalent to the Riemann hypothesis, and there are hundreds of statements that have been proven contingent on a future proof of the Riemann hypothesis.

Further Reading

Derbyshire, John. *Prime Obsession: Bernhard Riemann and the Greatest Unsolved Problem in Mathematics.* Washington, DC: Joseph Henry Press, 2003.

Matthews, Keith. "Number Theory Web: Biographies of Past Number Theorists and Various Items of Historical Interest." http://www.numbertheory.org/ntw/N14.html.

Oystein, Ore. *Number Theory and Its History.* New York: Dover, 1988.

Pommersheim, James, Tim Marks, and Erica Flapan. *Number Theory: A Lively Introduction with Proofs, Applications, and Stories.* Hoboken, NJ: Wiley, 2010.

Yan, Song. *Number Theory for Computing.* Berlin: Springer, 2002.

Michael "Cap" Khoury

for nonreal complex numbers was coined by René Descartes in 1637 to indicate that they do not really exist, a view later shared by Isaac Newton. About 1765, Leonhard Euler characterized square roots of negative numbers as impossible quantities, and as late as 1831, Augustus De Morgan objected to the absurd nature of complex as well as negative numbers. Only in 1837 did William Rowan Hamilton give a proper construction of the complex numbers, thereby indisputably proving their inner consistency. Nevertheless, it was their usefulness, the beauty of their simplicity, and the ability to visualize them rather than Hamilton's proof that eventually outweighed the objections against complex numbers and led to their universal acceptance by the end of the nineteenth century.

Algebra and Geometry of Complex Numbers

Complex numbers appeared for the first time in Gerolamo Cardano's *Ars Magna* from 1545. In this famous book containing the formulas for solving cubic and quartic equations, Cardano also showed that the equations $x + y = 10$ and $x \cdot y = 40$ have the common solution

$$x = 5 + \sqrt{-15} \text{ and } y = 5 - \sqrt{-15}.$$

Cardano, however, dismissed these complex numbers as useless and did not pursue the matter further. Rafael Bombelli undertook a more systematic investigation in *L'Algebra* from 1572, where he demonstrated how complex numbers can be added, subtracted, multiplied, and divided using the usual rules of algebra and the equation $i^2 = -1$. For example,

$$(1 + 3i) + (2 + i) = 3 + 4i$$

$$(1 + 3i) - (2 + i) = -1 + 2i$$

and $(1 + 3i) \cdot (2 + i) = 2 + i + 6i + 3i^2 = -1 + 7i$.

Division is slightly more complicated; it is most easily performed by multiplying both numerator and denominator by the conjugate of the denominator:

$$\frac{1+3i}{2+i} = \frac{(1+3i)\cdot(2-i)}{(2+i)\cdot(2-i)} = \frac{5+5i}{5} = 1+i.$$

Using these operations, Bombelli showed how real solutions to cubic equations can be found even when square roots of negative numbers appear in Cardano's formula for cubic equations. Bombelli's brilliant use of complex numbers for solving polynomial equations eventually led to the Fundamental Theorem of Algebra, according to which every polynomial equation of positive degree has a complex solution. The first essentially correct proof of this result, which had been anticipated already in the seventeenth century, was given by Carl Friedrich Gauss in 1799 in his doctoral dissertation.

Complex numbers can be represented geometrically as points in the complex plane, invented in 1797 by Caspar Wessel. Shortly afterward, it was independently conceived and popularized by Gauss, who used it implicitly in his proof of the Fundamental Theorem of Algebra. This concrete geometric interpretation of complex numbers was instrumental in the struggle to come to terms with their nature. In the complex plane, points on the x-axis correspond to real numbers, points on the y-axis to so-called purely imaginary numbers, and in general, the point with coordinates (a, b) corresponds to the complex number $a + bi$. Viewing complex numbers as points in the complex plane gives a new geometric understanding of Bombelli's rules of addition and multiplication. Also, the "numerical value" $|z|$ of a complex number $z = a + bi$ is defined geometrically as the distance between the points $(0, 0)$ and (a, b), or

$$|z| = \sqrt{a^2 + b^2}.$$

Machin's Formula and the Computation of π

John Machin, in 1706, discovered the formula

$$\frac{\pi}{4} = 4 \arctan\left(\frac{1}{5}\right) - \arctan\left(\frac{1}{239}\right)$$

and used it together with the Taylor series

$$\arctan(x) = x - \frac{x^3}{3} + \frac{x^5}{5} - \frac{x^7}{7} + \frac{x^9}{9} - \cdots$$

to compute π to 100 decimal places, a world record at the time. Although Machin's formula involves only real numbers, it has a surprisingly simple and elegant proof using the following identity of complex numbers, thus illustrating their utility in other areas of mathematics:

$$\frac{(5+i)^4}{239+i} = 2 \cdot (1+i).$$

Exponential and Trigonometric Functions

The exponential function e^x and the trigonometric functions $\cos(x)$ and $\sin(x)$ are well-known functions of a real variable x. They can be expressed as Taylor series as follows:

$$e^x = 1 + x + \frac{x^2}{2!} + \frac{x^3}{3!} + \frac{x^4}{4!} + \cdots$$

$$\cos(x) = 1 - \frac{x^2}{2!} + \frac{x^4}{4!} - \frac{x^6}{6!} + \frac{x^8}{8!} - \cdots \text{ and}$$

$$\sin(x) = x - \frac{x^3}{3!} + \frac{x^5}{5!} - \frac{x^7}{7!} + \frac{x^9}{9!} - \cdots.$$

Using these expressions, the complex functions e^z, $\cos(z)$, and $\sin(z)$ are defined for a complex variable z. With these definitions and the fundamental equation $i^2 = -1$, Euler, in 1748, proved a formula that reveals a surprising kinship between these seemingly unrelated functions:

$$e^{ix} = \cos(x) + i \cdot \sin(x).$$

This result, known as "Euler's formula," generalizes a formula found by Abraham de Moivre in 1730:

$$(\cos(x) + i \cdot \sin(x))^n = \cos(nx) + i \cdot \sin(nx).$$

Inserting $x = \pi$ into Euler's formula gives *Euler's identity*: $e^{i\pi} = -1$.

This identity combines the three most important mathematical constants—π, e, and i—into one single expression of striking simplicity and beauty. A 1988 poll of readers of *Mathematical Intelligencer* voted Euler's identity "the most beautiful theorem in mathematics," ahead of the infinitude of primes, the transcendence of π, and the Four-Color Theorem.

Complex Analysis

Complex analysis is the study of complex functions—functions $f(z)$ defined on some subset U of the set of complex numbers and with complex values. After initial contributions by Euler and Gauss, complex analysis was systematically investigated by Augustin-Louis Cauchy in the 1820s. Later in the nineteenth century, the theory was further developed by Bernhard Riemann and Karl Weierstrass. The set of definition U is called a "domain" if it is open and connected, and $f(z)$ is called "holomorphic" if it satisfies the condition of complex differentiability. Contrary to what the name suggests, complex analysis is in many ways simpler than real analysis, since complex differentiability is a much-stricter property than real differentiability. For example, every holomorphic function satisfies the so-called Cauchy-Riemann equations, which have no analogue in the realm of real functions. Also, the "Identity theorem" states that two holomorphic functions $f(z)$ and $g(z)$ defined on the same domain U are identical only if they agree on a line segment, a result very far from being true in real analysis. Other theorems and conjectures in complex analysis are concerned with other types of complex functions, such as entire and meromorphic functions.

An "entire" function is a holomorphic function defined on the entire complex plane. The complex exponential function e^z and the complex trigonometric functions $\cos(z)$ and $\sin(z)$ are examples of entire functions. Liouville's theorem, named after Joseph Liouville, states that every nonconstant entire function is unbounded. This theorem is considerably strengthened by Picard's little theorem, named after Charles Picard, which states that every nonconstant entire function takes every complex value with at most one exception. For example, $\cos(z)$ and $\sin(z)$ both take every complex value, whereas e^z takes every complex value except 0.

A meromorphic function is a quotient of two holomorphic functions defined on a domain U where the denominator is not identically zero. The zeros of the denominator are called "singularities"; they can be either "removable singularities" or "poles." The complex tangent function

$$\tan(z) = \frac{\sin(z)}{\cos(z)}$$

is an example of a meromorphic function; it has zeros at 0, $\pm\pi$, $\pm 2\pi$, and so on, and poles at $\pm\pi/2$, $\pm 3\pi/2$, $\pm 5\pi/2$, and so on. The mysterious Riemann zeta function $\zeta(z)$ is another example; it has a single pole at $z = 1$. The Riemann conjecture, arguably the most important unsolved problem in all of mathematics, states that all nonreal zeros of $\zeta(z)$ have real part equal to one-half. The Riemann conjecture is one of the seven Millennium Prize Problems for whose solution the Clay Mathematics Institute has offered a prize of $1 million.

Hamilton's Quaternions as Extensions of Complex Numbers

The complex numbers form an extension of the real numbers, just as the real numbers form an extension of the rational, integral, and natural numbers. It is therefore natural to ask if there are further numbers extending the complex numbers. This question was answered in the affirmative by Hamilton in 1843 when he discovered the quaternions. A quaternion is a number of the form $q = a + bi + cj + dk$ where a, b, c, and d are real numbers, and i, j, and k are symbols satisfying

$$i^2 = j^2 = k^2 = i \cdot j \cdot k = -1.$$

Quaternions, however, do not satisfy the commutative law of multiplication. For example, the product of i and j depends on the order of the factors: $i \times j \neq j \times i$. The numerical value of a quaternion q is defined as

$$|q| = \sqrt{a^2 + b^2 + c^2 + d^2}.$$

A quaternion with numerical value $|q| = 1$ is called a "unit quaternion." Each unit quaternion corresponds in a certain way to a rotation of three-dimensional space. For this reason, quaternions have important applications in computer graphics.

It happens that each unit quaternion q corresponds to the same rotation as its negative, $-q$. This mathematical subtlety explains one of the most surprising phenomena in quantum mechanics, namely, that the state of an electron is changed if the electron is rotated 360 degrees; only a rotation of 720 degrees leaves the electron unchanged.

Further Reading

Mazur, Barry. *Imagining Numbers (Particularly the Square Root of Minus Fifteen)*. New York: Farrar, Straus & Giroux, 2003.

Nahin, Paul J. *Dr. Euler's Fabulous Formula: Cures Many Mathematical Ills*. Princeton, NJ: Princeton University Press, 2006.

Stewart, I., and D. Tall. *Complex Analysis*. Cambridge, England: Cambridge University Press, 1983.

David Brink

Numbers, Rational and Irrational

Category: History and Development of Curricular Concepts.
Fields of Study: Communication; Connections; Number and Operations; Representations.
Summary: While the concept of rational numbers is easily understood, mathematicians has struggled with the concept of irrational numbers since antiquity.

Early philosophers and mathematicians explored whether real-life lengths were made up of whole numbers. The discovery of irrational numbers caused great concern and led to the development of number theory and real analysis. A rational number is a real number that can be written as a ratio of two integers. Real numbers that cannot be so written are called "irrational numbers." So, for example, 17/47 is a rational number, while π or $\sqrt{2}$ are irrational numbers. Rational and irrational numbers can also be represented using the decimal notation. The rational numbers are precisely those numbers whose decimal representation either terminates after a finite number of digits or is repeating. The decimal representation of irrational numbers does not have a repeating pattern. So, for example, 1/8 corresponds to 0.125 (a terminating decimal), while 1/7 corresponds to the repeating decimal 0.142857142857. . . . On the other hand, the irrational number π has a decimal expansion that begins with 3.14159265358979323846 . . . and continues indefinitely without any patterns.

Students in twenty-first-century classrooms explore rational numbers in middle school and irrational numbers in high school, and these numbers appear in nature and in daily calculations. For example, e, which is also irrational, is needed to calculate the interest compounded continually on a loan, and π appears in circular or spherical objects. In fact, as a consequence of Georg Cantor's work, given any real number, there is a higher probability of it being irrational. There are still open problems to explore, such as whether $\pi + e$ is irrational.

Definition

Irrational numbers are numbers that are not rational; in other words, any number that is not the ratio of two integers is an irrational number. This definition by itself, however, is circular. To be able to use it, one first has to know what a number is.

What is a number? This question is harder to answer than one might expect at first, and, in fact, has been contentious for most of the history of mathematics. The positive integers (or the counting numbers) 1, 2, 3, ... directly arise from the daily experience of humans, and it is impossible to trace how long ago humans went from the concrete ideas of three cows, three stones, and three trees and abstracted out the number 3 as a stand-alone concept. The advantage of this abstraction is that people could study operations on numbers and apply them to a large number of settings. If one knows that 3 + 5 = 8, this indicates simultaneously that 3 cows together with 5 cows are 8 cows, and that 3 trees and 5 trees are 8 trees. If one knows that 3 × 5 = 15, then, while 3 trees and 5 trees cannot be multiplied, this can be used to model many situations. Three boys each having five apples have a total of 15 apples, and a 3-by-5 piece of land has an area of 15. Mathematicians can now concentrate on finding better algorithms and methods for doing number operations. The concept of number was first enlarged to also encompass rational numbers—the ratios of positive integers.

Some 3500 years ago, Egyptians used unit fractions (reciprocals of positive integers) and 2/3 to pose and solve problems. For example, the third problem in the Rhind Mathematical Papyrus is about dividing 6 loaves among 10 men, and the answer given is

$$\frac{1}{2} + \frac{1}{10}.$$

Around the same time, Babylonian scribes in Mesopotamia used a base-60 place value system for fractions, but confined themselves to those rational numbers that have a finite sexagesimal representation. In any case, for a very long time, the term "number" meant positive integers and ratios of positive integers—what are now called the "positive rational numbers."

If the only numbers are rational numbers, then by definition there are no irrational numbers. To enlarge the definition of "number" beyond rational numbers, one has to somehow construct these other numbers, which can be done by proposing that the length of any line segment is a number.

It is believed that in early mathematics it was assumed that given any two line segments, it is possible to find a third line segment—maybe a very small one—that measures both lines a whole number of times. In other words, the length of each of the original line segments is an integer multiple of the third smaller line segment. The third line segment is called a "common unit of measure," and the original two line segments are called "commensurable." On the face of it, this assumption may seem reasonable, but, if true, it would mean that the length of any line segment is a rational number. Given an arbitrary line segment of length a, find a common unit of measure for it and a line segment of unit length. If the common unit of measure has length b, then $a = mb$ and $1 = nb$ for some integers m and n. But this means that

$$\frac{a}{m} = b = \frac{1}{n}$$

from which it can be determined that

$$a = \frac{m}{n}$$

is a rational number.

The Pythagorean Theorem, which was known at least 1000 years before Pythagoras, states that given a right triangle whose sides are of unit length, then the length of its hypotenuse will be such that yields 2 if multiplied by itself. Since it has been decided that all lengths are numbers, the length of this hypotenuse must be a number called the "square root of 2" and denoted by $\sqrt{2}$. One can prove that $\sqrt{2}$ is not a rational number, thereby proving that not all pairs of line segments are commensurable and that irrational numbers exist.

There are many proofs of the irrationality of $\sqrt{2}$, but the most common one is as follows: Assume by way of contradiction that $\sqrt{2}$ is rational and equal to n/m, where n and m are integers. One has many choices for n and m; for example, one could multiply both by 47 and get a new pair of integers with the same ratio—and values of n and m are chosen such that they do not have any common factors. This is, of course, possible. From $\sqrt{2} = n/m$, one obtains $2m^2 = n^2$, which means that n^2 and therefore n is an even number. If $n = 2k$, then $2m^2 = 4k^2$ and so $m^2 = 2k^2$, which means that m is also an even number. But it had been assumed that n and m have no common factors. A contradiction was reached but, since the logic along the way was impeccable, it must have been that the original assumption that $\sqrt{2}$ is rational must have been wrong.

Implications

The discovery of irrational numbers led to a crisis in geometry and a need to revisit all the results that depended on the commensurability assumption. Following Eudoxus, Euclid in his very influential book *Elements* makes a distinction between a number and a magnitude. Roughly, one can think of numbers as the rational numbers and magnitudes as the lengths of line segments—Euclid had an elaborate classification of magnitudes. In an attempt to be rigorous, Euclid treats number and magnitude differently, and hence he does not regard irrational numbers as numbers. For example, he develops the theory of proportions once for magnitudes and once for numbers.

It took the effort of many mathematicians in the middle ages—and most notably mathematicians living in Islamic lands and writing in Arabic—to expand the notion of number to include Euclid's magnitudes and to have a single treatment of all numbers, rational and irrational. During this period, the decimal number system—first developed in India and crucial in understanding irrational numbers—became widespread. Ninth-century Persian mathematician Al-Mahani gave a definition of irrational numbers (as opposed to Euclidean magnitudes), and ninth-century Egyptian mathematician Abu Kamil used irrational numbers as coefficients in algebraic equations. By the fifteenth century, Persian mathematician Jamshid Kashani (also referred to as al-Kashi) was able to comfortably work with real numbers and their decimal expansions. He treated both rational and irrational numbers as numbers.

In the West, sixteenth-century Flemish mathematician Simon Stevin played an important role in advocating the use of decimal fractions, in eliminating the Euclidean distinction between numbers and magnitudes, and in the understanding of real numbers as numbers. Finally, a modern rigorous construction and definition of real numbers (rational and irrational) was given by nineteenth-century German mathematician Richard Dedekind. He started with rational numbers and defined irrational numbers using the rational numbers.

Further Reading

Gouvea, Fernando. "From Numbers to Number Systems." In *The Princeton Companion to Mathematics*. Edited by Timothy Gowers. Princeton, NJ: Princeton University Press, 2008.

Katz, Victor J. *A History of Mathematics: An Introduction*, 2nd ed. Boston: Addison-Wesley, 1998.

Shahriar Shahriari

Numbers, Real

Category: History and Development of Curricular Concepts.
Fields of Study: Communication; Connections; Number and Operations.
Summary: The real number system is commonplace, but required centuries before it came to be understood in its modern form.

The real number system is often thought of as a number line, with each point on the line corresponding to a number. The set of real numbers includes all the integers and fractions (rational numbers); algebraic irrational numbers, such as the square root of 3 and the cube root of 19; and transcendental numbers such as π, $\log(2)$, and e, which do not satisfy any polynomial equation.

In the twenty-first century, students begin to explore whole numbers beginning in the earliest grades, as well as common fractions like 1/2 and 1/4. In the later primary grades they develop knowledge of base-10 decimal places, fractions as portions or divisions of a whole, negative numbers, and equivalent forms for fractions, decimals, and percentages. These notions are further expanded and applied in middle school, including concepts like ratios and proportions, integers, factorization, prime numbers, and exponential and scientific notation for very large numbers. Very large and small numbers, properties of numbers and various number systems, vectors and matrices with real number properties, and number theory may be studied in high school.

The real number system is the principal number system used in calculus, geometry, and measurement. In particular, when one uses coordinate (Cartesian) geometry to describe the plane or space, one labels points by pairs or triples of "real" numbers. In mathematics and the sciences, the word "number" without qualification is generally used to mean "real number."

Development of the Real Numbers

The ancient theory of length and measurement was very different from current understanding. The ancient Greeks (the civilization about which exists the most complete mathematical history) believed that any set of lengths were commensurable; in modern language, they believed that the ratio of any two lengths (or areas, or volumes) was a rational number. This was not a totally unreasonable belief, since indeed all lengths can be approximated very well by commensurable ones. It is not correct, though; for example, the ratio of the diagonal of a square to its side is the square root of 2.

Greek mathematician and numerologist Pythagoras knew this (it is a simple consequence of what is now called the "Pythagorean Theorem") and was further able to prove, contrary to the notion of commensurability, that no rational number, when squared, could equal 2. According to some stories, probably apocryphal, this discovery was so contrary to the belief system of Pythagoras and his followers that a discoverer was murdered or committed suicide. Ultimately, geometers were forced to accept the existence of irrational numbers.

The Greek mathematician and astronomer Eudoxus (c. 400–350 B.C.E.) wrote about the theory of proportions in a way that did not assume all lengths were commensurable and is generally credited with laying the groundwork for irrational numbers as legitimate mathematical objects.

Even after mathematicians realized that irrational numbers were required for practical purposes, the understanding of the real number line was somewhat vague and confused. Real numbers were understood, if at all, as things that could be approximated well by rational numbers or by decimal approximations. The major modern contribution to the understanding of real numbers was made by Richard Dedekind (1831–1916), who described the real numbers in terms of so-called "Dedekind cuts." In addition to its significance for abstract mathematics, Dedekind's insight also helped to explain some important phenomena in geometry (for example, why a line with points inside and outside a circle must intersect the circle).

This resistance to advancements in the understanding of number, this tendency for even very intelligent people to oppose enlarging the number system, even when doing so enables scientific and technological progress, is not unique to the ancient Greeks. A similar story unfolded much more recently with the development of the complex number system.

Decimal Representations

Every real number has a base-10, or decimal, representation. This consists of three components: a dot (called a "decimal point" in this context), a finite sequence of digits to the left of the decimal point (the integer part), and an infinite sequence of digits to the right (the fractional part). A digit can be 0, 1, 2, 3, 4, 5, 6, 7, 8, or 9. Working from the decimal point left, the digits occupy the ones place, the tens place, the hundreds place, the thousands place, and so on; from the decimal point right, the digits occupy the tenths place, the hundredths place, the thousandths place, and so on. In symbols, if the a_k are digits, then the decimal expansion

$$a_k a_{k-1} \cdots a_2 a_1 a_0 . a_{-1} a_{-2} \cdots a_{-k} \cdots$$

represents the real number

$$a_k 10^k + \cdots + a_1 10 + a_0 + a_{-1} 10^{-1} + \cdots + a_{-k} 10^{-k} + \cdots .$$

If from some point rightward, all the digits in the decimal representation of a number are zero, that expression is said to terminate, and the trailing zeroes are typically not written; for example, "7.24" instead of "7.24000...." Integers have only zeroes to the right of the decimal point, and in such cases even the decimal point is often omitted.

Relying exclusively on decimal expansions as a way to understand real numbers can be problematic. Specifying a real number in this way requires an infinite sequence of digits. Unless there is a pattern, this requires specifying an infinite amount of information. For example, there is no known digit-by-digit description for important numbers like π and e. Dealing with infinite expressions is confusing for many people. For example, some people find it difficult to accept that 0.33333... = 1/3, and even more people find it uncomfortable that 0.99999... = 1.

Almost all real numbers have a unique decimal expansion, but some have two. As 0.99999... = 1 illustrates, every number that can be written so that it ends in an infinite string of 0s also has an expansion that ends in an infinite string of 9s.

Structural Properties of the Real Number System

The real numbers form a field, which means that real numbers can be added, subtracted, multiplied, and divided (except by 0), and that the operations satisfy certain properties (for example, commutative, associative, and distributive laws). The real numbers are actually an ordered field, which means that there is a notion of what it means for one number to be less or greater than another that is compatible with the operations.

There is a natural way to measure distance between two numbers: the distance between numbers a and b is $|a-b|$, where $|\cdot|$ is the absolute value function. Loosely speaking, this means that one can talk about "closeness" of real numbers to each other; in technical language, the number line has a metric and a topology.

Unlike the set of integers (which is discrete), the real number line is continuous. The discrete/continuous distinction in mathematics is analogous to the digital/analog distinction in science and technology. A digital thermometer has discrete output, moving from 24 degrees to 25. An analog thermometer, on the other hand, can register 24 degrees or 24.65474 degrees or any other number. Unlike both the integers and the rational number system, the real number line is what called "topologically complete." Because of the ordering on the reals, this can be summarized as: "Any set of real numbers which has an upper bound has a least upper bound."

In mathematics history, adopting a continuous number system made it possible to develop "limits," the focal concept of calculus. The development of calculus, in turn, made possible numerous advances in sciences, especially physics and engineering.

Further Reading

Borwein, Jonathan, and Peter Borwein. *A Dictionary of Real Numbers*. Pacific Grove, CA: Brooks/Cole Publishing, 1990.

Burrill, Claude. *Foundations of Real Numbers*. New York: McGraw-Hill, 1967.

Drobat, Stefan. *Real Numbers*. Upper Saddle River, NJ: Prentice Hall, 1964.

Scriba, Christoph, with M. E. Dormer Ellis. *The Concept of Number: A Chapter in the History of Mathematics, with Applications of Interest to Teachers*. Zurich: Mannheim, 1968.

Stevenson, Frederick. *Exploring the Real Numbers*. Upper Saddle River, NJ: Prentice Hall, 2000.

Michael "Cap" Khoury

Parallel Postulate

Category: History and Development of Curricular Concepts.
Fields of Study: Communication; Connections; Geometry.
Summary: The parallel postulate led to thousands of years of investigation and debate.

One of humanity's greatest intellectual achievements occurred in approximately 300 B.C.E. when the axiomatic method was born. The classic text *Elements*, written by the great Greek geometer Euclid of Alexandria, is a work that shaped the nature of mathematics and stands to this day as an example of the beauty and elegance of reasoning and proof.

Euclid was among the first people to understand that abstract mathematics is based on reasoning, from assumptions to general conclusions. From a very modest set of assumptions—his five postulates (called "axioms")—Euclid set out to argue the truth of a large number of propositions (called "theorems") in geometry.

The first four of Euclid's postulates appear reasonable enough: (1) any two points determine a unique line; (2) any line segment can be extended to an infinite line; (3) given any center and radius, a circle can be constructed; and (4) all right angles are congruent. But the fifth postulate stands out for its comparative complexity:

If a straight line falling on two straight lines makes the interior angles on the same side less than two right angles, the two straight lines, if produced indefinitely, meet on that side on which are the angles less than the two right angles.

This fifth postulate has come to be known as the "parallel postulate," in part for its very content, but also for the key role it plays in proving certain propositions about parallel lines.

From an historical perspective, Euclid himself seemed a bit uncomfortable with his fifth postulate.

This discomfort is evidenced by the order of his work in Book I of *Elements*, where, on his way to eventually proving 48 propositions, he waited until proposition 29 to use the parallel postulate. The first 28 results rely only on the first four postulates and theorems that can be proven using those assumptions.

Attempts to Prove the Parallel Postulate as a Theorem

As subsequent mathematicians studied the *Elements*, most were troubled in some way by the parallel postulate. Because of its complexity, as well as its "if-then" format, it struck most mathematicians that Euclid's fifth postulate really ought to be a theorem. In other words, the parallel postulate ought to be a consequence of the first four postulates, and this fact ought to be provable, using only those four postulates and any theorems that could be derived from them.

Thus, many mathematicians set out to prove the parallel postulate as a theorem. It is one of the great tales of the history of mathematics that every single mathematician who attempted to prove the parallel postulate failed. Early on, many of these esteemed intellects made a common error that the rules of logic forbid—they assumed precisely what they were attempting to prove. Clearly, if the goal is to prove a statement S, one should never be allowed to simply assume that S is true. While certainly no mathematician was so dull as to say, "To prove the parallel postulate, I will assume the parallel postulate," many people did make the mistake of making the assumption that certain "obvious" statements were true. For example, they may have assumed statements such as the following:

- Parallel lines are everywhere equidistant.
- The sum of the measures of the interior angles of a triangle is 180 degrees.
- If a line intersects one of two parallel lines, then it must also intersect the other.
- There exists a rectangle (a quadrilateral having four right angles).

Remarkably, each of the above statements (along with many others) is equivalent to the parallel postulate. Said differently, if one of the above statements is called P and the statement of the parallel postulate is called S, then it turns out that P is true if and only if S is true—the truth of one implies the truth of the other, and vice versa.

Hence, when a mathematician said, "Using the fact that any triangle's angle sum is 180 degrees," and then went on to "prove" the parallel postulate, this argument was like saying "the parallel postulate is true because the parallel postulate is true." These errors came to be well understood by the end of the eighteenth century, perhaps most prominently in G. S. Klugel's 1763 doctoral dissertation in which he debunked 43 flawed "proofs" of the parallel postulate.

Modern Conceptions

Today, mathematicians understand a great deal about the role of Euclid's parallel postulate. Euclid's parallel postulate really is an axiom, and not a theorem. The parallel postulate is independent of the first four postulates. One can assume that the parallel postulate is true, or one can assume that the parallel postulate is false. Either leads to a perfectly valid geometry, with the truth of the parallel postulate leading to Euclidean geometry. Considering Playfair's postulate, named for John Playfair, if one assumes there are no parallel lines through a point P not on a line l, then this leads to so-called elliptic geometry, which is like the geometry of the sphere. If instead one assumes that there is more than one parallel line through a point P not on l, then this leads to "hyperbolic geometry," a geometry that some believe may help describe the shape of the universe.

It took approximately 2000 years for humankind to fully appreciate the work of Euclid and to reconcile the fact that Euclid was right—his fifth postulate really is an axiom, and not a theorem that can be derived. More than this, the parallel postulate is like a door that opens the world to one geometry—Euclidean—while there are other similar postulates that open doors to different universes, those of elliptic and hyperbolic geometries.

Girolamo Saccheri's Developments

Of course, even though nobody had found a valid proof of the parallel postulate did not mean that one could not be found, and many continued the search. Around the turn of the eighteenth century, a Jesuit priest named Girolamo Saccheri (1677–1733) made a lasting contribution to the study of the parallel postulate in particular, and to the history of mathematics in general. Saccheri considered the unthinkable, as part of his effort to prove the parallel postulate through a contradiction argument: what if the parallel postulate is false?

It was well understood by Saccheri's time that an equivalent statement of the parallel postulate was Playfair's Postulate, which states that

For any line l and any point P not on l, there exists a unique line through P parallel to l.

A contradiction argument works by assuming that the statement one wants to prove true is actually false and showing that some contradiction follows. Thus, it is natural to consider Playfair's Postulate and suppose that there is not be a unique line through P parallel to l. That is, one would assume that either there is not *any* line through P parallel to l, or there is *more than one* line through P parallel to l. Saccheri considered a similar scenario where he had transformed the problem about parallels to an equivalent one about quadrilaterals (now called "Saccheri quadrilaterals") in which the quadrilateral has two congruent sides perpendicular to the base. Fundamentally, Saccheri was trying to prove that a rectangle existed by showing that the summit angles of his quadrilateral were also right angles. After proving that the summit angles were congruent, he realized that there were three possibilities: the summit angles were each right angles, each was less than a right angle, or each was more than a right angle.

While Saccheri was able to rule out the possibility that the summit angles were obtuse by assuming that they were obtuse and finding a contradiction, when he assumed that the summit angles were acute, he could not find a contradiction. From this assumption, he went on to prove many strange and unusual theorems. Unknowingly, Saccheri had discovered a whole new geometry, one that another mathematician named Janos Bolyai would call "a strange, new universe" in his own investigations. What both of these mathematicians, along with others such as Carl Gauss, started to realize is that there actually exists a geometry in which there is more than one line through a point P not on line l such that each is parallel to l. This realization stands as one of the greatest accidental discoveries in the history of the human intellect: Saccheri did not find what he set out to prove, but instead developed a collection of ideas that would radically change mathematics.

Further Reading

Dunham, Douglas. "A Tale Both Shocking and Hyperbolic." *Math Horizons* 10 (April 2003).

Greenberg M. *Euclidean and Non-Euclidean Geometries: Development and History*. New York: W. H. Freeman and Co., 2007.

Socrates Bardi, Jason. *The Fifth Postulate: How Unraveling A Two Thousand Year Old Mystery Unraveled the Universe*. Hoboken, NJ: Wiley, 2008.

Matt Boelkins

Perimeter and Circumference

Category: History and Development of Curricular Concepts.
Fields of Study: Communication; Connections; Geometry; Measurement.
Summary: Measuring perimeter and circumference is a geometric task with a long history of methods.

Measurements of length and distance abound in daily life, from the height of a child to the distance from home to the store. Perimeter and circumference are types of length measurements. The perimeter of a geometric entity is the path that surrounds its area. The word derives from its Greek roots *peri* (meaning "around") and from *meter* (meaning "measure"). In stricter mathematical sense, perimeter is defined as the length of the curve constituting the boundary of a two-dimensional, planar closed surface.

For example, the perimeter of a square whose side measures length a is $4a$. Perimeter is important for applications such as landscaping projects, construction, and building fences. Circumference is defined as the perimeter of a circle. The circumference of a circle of radius (r) is $2\pi r$. The circumference of a circle has played a very important role throughout history in the approximation of the mathematical constant π, which was defined as the ratio of the circumference (C) of the circle to its diameter (d). Perhaps the most common reference to circumference that most people encounter regularly is the circumference of one's waist—the size of their waist.

The waist circumference is used as a measurement for some clothing and is also associated with type II diabetes, dyslipidemia, hypertension, and other cardiovascular diseases. Students investigate perimeter beginning in primary school, and middle grade students explore circumference. Students formulate the length of general curves, referred to as the "arc length" or "rectification," as integrals in calculus courses.

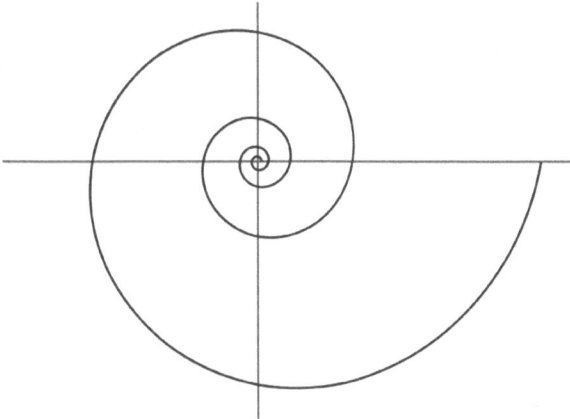

The curve that forms the shape of a nautilus shell is known as a logarithmic spiral or equiangular spiral.

History

There is a long history of computations involving the perimeter or circumference of figures. One way was to measure length was with ropes. For instance, statements about rope measurements and the Pythagorean theorem can be found in Katyayana's *Sulbasutra*. Another way was to compare the length of two figures. A Babylonian clay tablet was discovered in 1936 and was noted as relating the hexagon perimeter to 0;57,36 (in base 60), or 24/25 times the circumference of a circumscribed circle. Mathematicians like Archimedes of Syracuse estimated the circumference or a value for π by using the perimeters of inscribed and circumscribed polygons with many sides. For example, Archimedes was known to have used 96-sided polygons. Mahavira estimated the circumference of an ellipse. In ancient times, the semiperimeter, or half the perimeter, was useful in computing many geometrical properties of polygons such as altitude, exradius, and inradius of a triangle. The semiperimeter also appears in Heron of Alexandria's formula for the area of a triangle. The semiperimeter of a rectangle is the sum of the length plus the height and is noted as appearing on Babylonian clay tablets. Brahmagupta used the semiperimeter of a quadrilaterial in the computation of its area.

The circle is a special geometric figure, for it is the curve, given a fixed perimeter, which encompasses the maximum surface area. This is known as the isoperimetric problem. Proclus commented that, "a misconception is held by geographers who infer the size of a city from the length of its walls." The Babylonians may have worked on related problems in their investigations of solutions to quadratic equations generated by the setting of the semiperimeter and area to constants. The isoperimetric problem was partially solved by the Greek mathematician Zenodorus.

Pappus of Alexandria compared the areas of figures with a fixed perimeter. In the tenth century, Abu Jafar al-Khazin proved that an equilateral triangle has greater area than isosceles or scalene triangles of the same fixed perimeter. Many mathematicians worked on the isoperimetric problem using a variety of techniques including methods from geometry, analysis, vectors, and calculus. In 1842, a German mathematician named Jakob Steiner used geometric arguments to present five proofs of the theorem. However, Steiner had assumed that a solution was possible, which was a subtle flaw to otherwise creative arguments. Karl Weierstrass proved the existence of such solutions in 1879. Other mathematicians proved the results in a variety of other ways.

Historical Applications and Computations

One application of circumference of a circle is the computation of the Earth's circumference. Eratosthenes of Cyrene, in 240 B.C.E., computed the Earth's circumference using trigonometry and the angle of elevation of the sun at noon in Alexandria and Syene. He made an assumption that the Earth and the sun were perfect spheres and that the sun was so far away that its rays hitting the Earth could be considered parallel. By measuring the shadows thrown by sticks on the summer solstice, Eratosthenes derived a formula to measure the circumference of the Earth and determined it to be 252,000 stadia. Teachers in mathematics classrooms share Eratosthenes's calculation in order to highlight his ingenuity and showcase the power of setting up proportions and applying the congruence of alternate interior angles of parallel lines. There is debate about the value of a stadia, but historians estimate that Eratosthenes was correct within a 2% to 15% margin of error. The Indian mathematician Aryabhata made revolutionary contributions toward the understanding of astronomy at the turn of the fifth century. His calculations on π, the circumference of Earth, and the length of the solar day were remarkably close approximations.

The middle of the seventeenth century marked a fruitful time in the history of calculating the length of general curves. For instance, the curve that forms the shape of a nautilus shell is called the "logarithmic spiral" or "equiangular spiral." Evangelista Torricelli described its length using geometric methods. Christopher Wren published the rectification of the cycloid curve. Hendrik van Heuraet and Pierre de Fermat independently explored ideas that would eventually lead to the integral formula of arc length.

In the twentieth century, methods from fractals, popularized by Benoit Mandelbrot, have proven useful in modeling objects like a coastline. One example that is regularly examined in mathematics classrooms is the Koch snowflake, named for Helge von Koch, an example of a curve that bounds a region with finite area yet has infinite perimeter.

Further Reading

Blasjo, Viktor. "The Isoperimetric Problem." *The American Mathematical Monthly* 112, no. 6 (2005).

Briggs, William. "Lessons From the Greeks and Computers." *Mathematics Magazine* 55, no. 1 (1982).

Dunham, William. "Heron's Formula for Triangular Area." In *Journey through Genius: The Great Theorems of Mathematics*. Hoboken, NJ: Wiley, 1990.

Steinhaus, H. *Mathematical Snapshots*. 3rd ed. New York: Dover, 1999.

Wells, D. *The Penguin Dictionary of Curious and Interesting Geometry*. London: Penguin, 1991.

Ashwin Mudigonda

Permutations and Combinations

Category: History and Development of Curricular Concepts.
Fields of Study: Communication; Connections; Data Analysis and Probability; Number and Operations.
Summary: For centuries, mathematicians have posed and studied problems that involve various arrangements or groupings of sets of objects, which are known as permutations and combinations.

In a very broad sense, combinatorics is about counting. The mathematical discipline of combinatorics addresses the enumeration, permutation, and combination of sets of objects, as well as their relations and properties. Combinatorial problems can be found in many areas of pure and applied mathematics, including algebra, topology, geometry, probability, graph theory, optimization, computer science, and statistical physics. One of the earliest problems in combinatorics is found in the work of Greek biographer Plutarch, who described mathematician Xenocrates of Chalcedon's work on calculating how many syllables could be produced by taking combinations of the letters of the alphabet. This occurred between 400 and 300 B.C.E. Millennia later, mathematician William Gowers won the 1998 Fields Medal, widely regarded as the most prestigious prize in mathematics, for his "contributions to functional analysis, making extensive use of methods from combination theory." In twenty-first-century school curricula, primary school children study number combinations to facilitate learning basic operations like addition, subtraction, multiplication, and division. High school students often study permutations and combinations as counting techniques. Permutations and combinations were fundamental for cracking the World War II Enigma code and continue to remain vital in cryptography, among other fields.

Definitions

In mathematical fields like algebraic group theory, combinatorics, or probability, the term "permutation" has several meanings that are all essentially related to the idea of rearranging, ordering, or permuting some kind of mathematical object. When paired with combinations, particularly in primary and secondary curricula, a permutation is usually thought of as an ordered arrangement of some set or subset of objects. For example, for the set of objects A, B, and C, there are six permutations of the set: {A, B, C}, {A, C, B}, {B, A, C}, {B, C, A}, {C, A, B}, and {C, B, A}. A combination is then a subset of objects selected from a larger set, where order does not matter. For example, for the set A, B, and C, one combination of two objects is {A, B}. In some applications, the objects in a combination are thought of as being chosen sequentially. However, since order does not matter, the selection {B, A} would represent the same combination as {A, B}. All possible two-object combinations are {A, B}, {B, C}, {A, C}. Mathematicians Blaise Pascal and Gottfried Leibniz used the specific term "combinations" beginning in the seventeenth century, while Jacob Bernoulli is often credited with introducing the term "permutations" a short while later. Some alternatively trace it to Thomas Strode in the seventeenth century.

History and Early Applications

The real-world motivation for many early problems involving what are now called "combinations" and "permutations" was religion. For example, Jaina, Christian, and Jewish scholars were interested in letter permutations, which some believed had spiritual power. In the ninth century, the Jaina mathematician Mahavira discussed rules for using permutations and combinations. In the tenth century, Rabbi Abraham ben Meir ibn Ezra used combinations to study the conjunction of planets. Another motivator was games of chance, which also drove probability theory. Archaeological evidence suggests that gambling has been around since the dawn of humankind, and many games rely on players achieving special combinations of symbols or objects like knucklebones, sticks, or polyhedral dice. Surviving writings show that Egyptian, Greek, Hindu, Islamic, and perhaps Chinese scholars and mathematicians studied permutations and combinations.

The Egyptian game "Hounds and Jackals" used a set of "throw" sticks that resulted in combinations of outcomes that determined how far a player might move. In the sixth century B.C.E., Hindus discussed combinations of six tastes: sweet, acid, saline, pungent, bitter, and astringent. Some consider the Chinese divination text *I-Ching* to be part of the literature on combinations and permutations since it discussed arrangements sets of trigram and hexagram symbols. Versions date to at least 400–300 B.C.E. In the sixth century, Roman philosopher Anicius Manlius Severinus Boëthius presented a rule for finding the possible combinations of objects taken two at a time from some set.

In the tenth and eleventh centuries, mathematicians like Acharya Hemachandra explored the how many combinations of short and long syllables were possible in a line of text with a fixed length, and Bhaskara's treatise *Bhaskaracharyai* contained an entire chapter devoted to combinations, among other chapters on topics like arithmetic, geometry, and progressions. Both al-Marrakushi ibn Al-Banna and Kamal al-Din Abu'l Hasan Muhammad Al-Farisi explored the relationship between polygonal numbers, the binomial theorem, and combinations. Al-Farisi used what historians consider a form of induction to show the relationship between triangular numbers (numbers that can be represented by an equilateral triangular grid of points such as, 1, 3, 6, 10), and the combinations of subsets of objects drawn from a larger set. Mi'yar al-'aqul ibn Sina (Avicenna) developed a system of combinations of "simple" machines to classify complex mechanisms.

The original concept of simple machines is attributed to mathematician Archimedes of Syracuse. A group might be machines containing rollers and levers, chosen from a larger set of possibilities that included windlasses, pulleys, rollers, levers, and other components. Starting in the Renaissance, the most commonly recognized set of six simple machines was the lever, inclined plane, wheel and axle, screw, wedge, and pulley. Students continue to discuss more complex machines as combinations of simple machines.

In Europe, beginning around the twelfth century and up through the nineteenth century, many mathematicians such as Levi ben Gerson, Bernoulli, Leibniz, Pascal, Pierre Fermat, Abraham de Moivre, George Boole, and John Venn worked on the development of combinations and permutations, frequently in the context of probability theory. For example, Johann Buteo (or Jean Borell) discussed the possible throws of four dice as well as locks with movable combination cylinders in his sixteenth-century work *Logistica*.

Bernoulli's *Ars Conjectandi* collected knowledge of permutations and combinations through the seventeenth century and was a popular combinatorics book in the eighteenth century. However, standard notation for permutations and combinations was still emerging.

Factorials

A mathematical function called a *factorial* is used to compute the number of possible permutations and combinations. Let $n!$ equal

$$n \times (n-1) \times (n-2) \times (n-3) \times \ldots \times 3 \times 2 \times 1.$$

For example, $5! = 5 \times 4 \times 3 \times 2 \times 1 = 120$. Further, $0!$ is defined to be 1. Bernoulli had proved many factorial results, like the fact that $n!$ gives the number of permutations of n objects. The use of the exclamation point to indicate a factorial, which was more convenient for printers of the day than some older notations, has been attributed to mathematician Christian Kramp. He worked in the late eighteenth and early nineteenth centuries. The general rule for finding permutations and combinations is sometimes attributed to Bernoulli and sometimes to sixteenth and seventeenth century mathematician Pierre Hérigone, who is also famed for introducing a variety of mathematical and logical notations. However, mathematicians used their own methods for indicating permutations and combinations well into the nineteenth century. For example, Thomas Harriot's seventeenth-century work *Ars Analyticae Praxis* contained unique symbolism for displaying the combinatorial process of finding binomial products.

In the notation common in the twentieth and twenty-first centuries, the number of permutations is stated as $n\mathrm{P}r$ where n is the total number of objects in a set and r is the number of objects selected from n and permuted,

$$n\mathrm{P}r = \frac{n!}{(n-r)!}.$$

The number of combinations is $n\mathrm{C}r$, which is read as "n choose r,"

$$n\mathrm{C}r = \binom{n}{r} = \frac{n!}{r!(n-r)!}.$$

The partial origins of this approach may perhaps be traced to nineteenth-century amateur mathematician Jean Argand, who used (m, n) to represent combinations of n objects chosen from a set of m objects.

Modern Developments

In the early twentieth century, mathematicians and others continued to develop theories and applications of combinatorial concepts. For example, statistician Ronald Fisher applied combinations to the design of factorial experiments, while artist Maurits Cornelius (M.C.) Escher developed his own system for categorizing combinations of shape, color, and symmetrical properties, which can be found in his 1941 notebook later referred to as a paper, *Regular Division of the Plane with Asymmetric Congruent Polygons*. Historians discuss that the sketchbooks of a typical artist contain preliminary versions of final works. Escher's book, on the other hand, appeared to form a theoretical mathematical basis for his tiling work. These combinatorial categories also influenced the field of crystallography.

Circular permutations are also common. One could think of lining up six people in a straight line to take their picture versus seating them at a round table. There are $n!$ permutations of the people lined up. However, once all six people are seated, even if they were all asked to move over one seat, they would all still be seated in the same overall order. There are therefore $(n-1)!$ ways of putting objects in a circle. Another possibility is that all items in the set are not unique, like the letters in "Mississippi," which reduces the number of unique permutations and combinations versus a set of the same length with unique components.

Permutation Groups

In a field like modern algebra, permutations can be viewed as maps that relate a set to itself. The set of permutations is then collected into an algebraic structure called a "group." One example is the various possible transformations of a Rubik's Cube puzzle, named for Erno Rubik. There are 43,252,003,274,489,856,000 permutations in the group for a 3-by-3-by-3 Rubik's Cube. Mathematicians often use software like the Groups, Algorithms, Programming (GAP) system to model and understand the transformations. Theories about permutation groups have been traced by historians to at least as far back as Joseph Lagrange's 1770 work *Réflexions sur la résolution algébrique des équations*, in which he discussed the permutations of the roots of equations and considered those roots as abstract structures. Paolo Ruffini used what would now be called *group theory* in his work, including permutation groups, and proved many fundamental theorems. In the nineteenth century, Augustin-Louis Cauchy generalized some of Ruffini's results. He studied permutation groups and proved what is now known as Cauchy's theorem. High school mathematics teacher Peter Sylow wrote his book *Théorèmes sur les groupes de substitutions* in the latter half of the nineteenth century, and it contained what are now known as the three Sylow theorems, which he proved for permutation groups. Arthur Cayley wrote about the connections between his work on permutations and Cauchy's, extended the notion of permutation groups into the broader idea of algebraic groups, and ultimately proposed that matrices and quaternions were types of groups. Some of his work served as one foundation for physicist Werner Heisenberg's development of quantum mechanics.

In the early twentieth century, George Pólya used permutation groups and other methods to enumerate isomers (compounds that have the same molecular components but different structural arrangements, or permutations) in organic chemistry. He also influenced Escher's studies of combinations. The George Pólya Prize is given every two years by the Society for Industrial and Applied Mathematics. One criterion for winning is "a notable application of combinatorial theory." Mathematicians continue to explore permutations and combination concepts in algebra and many other areas of mathematics.

Further Reading

David, F. N. *Games, Gods & Gambling: A History of Probability and Statistical Ideas*. New York: Dover Publications, 1998.

Davis, Tom. "Permutation Groups." http://www.geometer.org/mathcircles/perm.pdf.

Higgins, Peter. *Number Story: From Counting to Cryptography*. New York: Copernicus, 2008.

Carmen M. Latterell

Pi

Category: History and Development of Curricular Concepts.
Fields of Study: Communication; Connections; Measurement; Geometry.
Summary: The ratio of a circle's circumference to its diameter, π, is one of the most important constants and the first irrational number encountered by most students.

By definition, pi (π) is the ratio of a circle's circumference to the diameter. This definition holds for any circle, with the value of π being the constant value 3.14159265358979.... This decimal neither terminates nor repeats, making π irrational. Mathematicians and non-mathematicians alike are intrigued by the many appearances of π in diverse situations. Capturing this apparent mysticism in the 1800s, the mathematician Augustus de Morgan wrote, "This mysterious 3.14159… which comes in at every door and window, and down every chimney."

Values Used for Pi

Since the beginning of written mathematics, people have tried to calculate π's value. Around 2000 b.c.e., the Babylonians and Egyptians assigned values equal to 3 1/8 (3.125) and $4(8/9)^2$ (3.1605). In 1100 b.c.e., the Chinese used π=3, a value which also appears in the Bible (I Kings 5:23). In 300 b.c.e., Archimedes of Syracuse produced the first "accurate" value, using inscribed and circumscribed 96-sided polygons to produce the approximation 3 10/71 < π < 3 1/7 (or 3.140845... < π < 3.142857...). Since that time, multiple methods and formulas have been created to determine more exact values of π. Today, powerful computers use similar formulas to calculate values of π to extreme precision, with the current value exceeding 2.7 trillion digits (the record as of January 2010). Two examples of these formulas involving infinite series are

$$\frac{\pi}{2} = \frac{2 \times 2 \times 4 \times 4 \times 6 \times 6 \times 8 \cdots}{1 \times 3 \times 3 \times 5 \times 5 \times 7 \times 7 \cdots}$$

or $\frac{\pi}{4} = \frac{1}{1} - \frac{1}{3} + \frac{1}{5} - \frac{1}{7} + \frac{1}{9} - \cdots$.

Students in the twenty-first century learn about π in elementary school, and exposure to π continues in later courses in mathematics and physics. Since spherical coordinates are used in many applications, π is found in physical formulas such as Einstein's field equations, the Heisenberg uncertainty principle, and Coulomb's law for electric force, which are named after Albert Einstein, Werner Heisenberg, and Charles-Augustin de Coulomb, respectively. Mathematicians and computer scientists describe π as a great stress test for computers because of the seemingly random aspects of its digits.

Algorithms to compute the digits of π are regarded as more important than the digits themselves. Mathematicians continue to investigate other unsolved problems related to π, including attempts to determine how random the digits are.

Applications

The number π has played important roles in multiple situations. In 1767, Johann Lambert proved that π was irrational (it could not be written as the ratio of two integers). Then, in 1882, Ferdinand von Lindemann proved that π was transcendental (it could not be constructed using geometric tools and was not a root of a non-constant polynomial equation with rational coefficients). These two discoveries provided the key to proving the impossibilities of the Greeks' three problems of antiquity—squaring a circle, trisecting an angle, and duplicating a square.

Considered by many to be a ubiquitous number, π shows up in odd situations. First, in 1777, the naturalist Georges Buffon approximated the value of π experimentally by tossing a needle (length L) on a ruled surface (parallel lines spaced at distance D). If the tossed needle touches a line S times on N tosses, then

$$\pi \approx \frac{2SL}{DN}.$$

Second, the probability that two random integers are relatively prime (they have no common divisor) is

$$\frac{6}{\pi^2}.$$

Anyone can try these experiments, either by dropping needles or taking ratios of random integers; many are surprised that both produce good approximations for π. However, complex mathematics is needed to explain "why."

In 1743, Swiss mathematician Leonhard Euler published the formula $e^{ix} = \cos(x) + i\sin(x)$, linking exponentials, trigonometric functions, and complex numbers. Substituting $x=\pi$, the result becomes the most beautiful formula in mathematics: $e^{i\pi}+1=0$.

Popular Culture

The fascination with decimal expressions of π has led to competitive memorization contests. The Guinness World Records officially recognized Lu Chao as the most recent record holder in the early twenty-first century, but others have claimed more digits. Some people use piems (mnemonic poems); for example, "How I need a drink, alcoholic of course, after the heavy lectures involving quantum mechanics." In this piem, replace each word with its number of letters, producing $\pi \approx 3.14159265358979$. Hideaki Tomoyori, who held the world record of 40,000 digits memorized from 1987–1995, used a pictorial mnemonic system and explained, "I want to go on with the challenge of memorizing π, for just the same reason that people climb high mountains. I think it's a wonderful thing to challenge the limits of what we can do.... the more one memorizes of it, the closer one comes to the real value of the circle—closer to perfection." Researchers compared his cognitive abilities with a control group and concluded that he was not superior; they attributed his achievement to extensive practice.

The number π also is connected to some odd events. In 1897, the Indiana State Legislature almost passed a mathematically incorrect bill relating to π and squaring the circle. By its definition, the value of π changes if the circle shifts out of the Euclidean world. That is, in taxicab geometry, or metric geometry on a rectangular lattice structure, the value of π is 4.

The number π is an amazing number, both in its interesting properties and the obsessive attention given it by both mathematicians and non-mathematicians. How else could one explain why on March 14 at 1:59, many people shout, "Happy Pi Day!"

Further Reading

Adrian, Y. E. O. *The Pleasures of Pi, e and Other Interesting Numbers*. Singapore: World Scientific Publishing, 2006.

Beckmann, Petr. *A History of π (Pi)*. New York: Barnes & Noble, 1971.

Berggren, Lennart, Jon Borwein, and Peter Borwein. *Pi: A Source Book*. New York: Springer-Verlag, 1997.

Blatner, David. *The Joy of π*. New York: Walker & Co., 1997.

Takahashi, Masanobu, et al. "One% Ability and Ninety-Nine% Perspiration: A Study of a Japanese Memorist." *Journal of Experimental Psychology. Learning, Memory, and Cognition* 32, no. 5 (2006).

Jerry Johnson

Polygons

Category: History and Development of Curricular Concepts.
Fields of Study: Communication; Connections; Geometry.
Summary: Polygons have properties making them important in engineering, architecture, and elsewhere.

Shapes and figures define how people view the world. Polygons are special figures whose properties and relationships are prevalent in nature and are used extensively by architects, engineers, scientists, landscapers, and artists. Specifically, polygons are traditionally planar (two-dimensional) figures that are closed and comprised of line segments that do not cross. These line segments are called "edges" or "sides," and the points where the edges meet are called "vertices." Planar polygons are very important in engineering, computer graphics, and analysis because they are rigid, they work well with functions, and they are easy to transform. Other types of polygons are also useful, such as spherical, hyperbolic, complex, or near polygons.

Properties of Polygons

Polygons are named by the number of their sides. Typically, polygons with more than 10 sides are called n-gons.

Calculating angle sums, areas, and perimeters of polygons is important in architecture, landscaping, and interior design. Understanding properties of triangles and parallelograms facilitates these kinds of calculations. For instance, the sum of the measures of the interior angles of a polygon can be determined by realizing that a polygon with n sides can be divided into $n-2$ triangles, and that the sum of the measures of the interior angles of any triangle is 180 degrees. Using these ideas, a carpenter could easily determine the angles at which, for example, the sides of a hexagonal window frame should meet. Furthermore, the ability to create polygons from triangles and the ability to rearrange or duplicate some polygons to form parallelograms allow the derivation of area formulas. Michael Serra describes in his 2008 book, *Discovering Geometry: An Investigative Approach*, how the area of a parallelogram can be derived from a rectangle, and the area of a triangle can be derived from a parallelogram.

Real World Examples

Polygons are prevalent in the world. Even traffic signs come in the shapes of triangles, rectangles, squares, kites, and octagons. The properties of polygons make them useful in many areas including architecture, structural engineering, nature, and art.

Polygons are sometimes used in architecture for their structural benefits. Trusses formed from triangles provide support for bridges and roofs because, unlike other polygons, triangles do not tend to deform when force is exerted on a vertex. Fences are often formed into polygons because they can be built by linking together straight segments of material that are of equal size and shape. The buildings that comprise the Pentagon building in Washington, D.C., are arranged in a pentagonal shape because, according to Stephen Vogel, walking distances between buildings are less than in a rectangle, straight sides are easier to build, and the symmetrical shape is appealing. In the 1850s, Orsen Fowler popularized octagonal-shaped houses because octagons have larger areas than rectangles with the same perimeter. Thus, octagonal houses provided maximal living space while keeping heating, cooling, and building costs similar to that of the smaller rectangular house with the same outer wall space.

Properties of quadrilaterals and triangles facilitate the creation of squares and right angles. For example, using the properties of a square's diagonals, an approximate baseball diamond could be constructed by cutting diagonals of equal length from string or rope. To form the square, the diagonals would be positioned to bisect (halve) each other at right angles. The ends of each string would then mark the square's four corners. The same format could be used to create a rectangular play area, except the diagonals would not be perpendicular. According to Sidney Kolpas, although unaware of the Pythagorean theorem, ancient Egyptians used right triangles to reconstruct property boundaries after the annual flooding of the Nile River. To create a 90 degree angle, Egyptians would create a 3-4-5 right triangle by tying 13 equally spaced knots in a rope, placing stakes at knots 4 and 8, then drawing the ends of the rope at knots 1 and 13 to meet.

Polygons are prevalent in nature. Mineral crystals often have faces that are triangular, square, or hexagonal. The cross section of the Starfruit is shaped like a pentagonal star. Katrena Wells describes practical applications of hexagons, such as the often hexagonal shape of snowflakes and the hexagonal markings on many turtles' backs.

Tessellations of polygons are arrangements of polygons on a plane with no gaps or overlaps. These are also seen frequently in nature. Marvin Harrell and Linda Fosnaugh discuss many examples, including the facts that bees use a hexagonal tessellation for their honeycomb, some plant cell structures form hexagonal tessellations, and cooling lava may have formed the tessellating hexagonal columns of basalt rock at the Giant's Causeway in Ireland. Interestingly, a giraffe's skin is covered with a tessellation of various approximate polygons.

When creating sketches of objects or animals, artists often use polygons as the basis of their work by breaking the figure down into polygons and circles, then smoothing and filling in the details of the drawing after the rough polygonal sketch is created. Michael Serra explains how artist M.C. Escher used tessellations of triangles, squares, and hexagons as a framework, then rotated or translated various drawings along the sides of each polygon in the tessellation to create marvelous patterns of reptiles, birds, and fish. Islamic artists covered their buildings with ornate tessellations of polygons. A prime example is the Alhambra Palace in Grenada, Spain.

Investigating polygons as they exist in the world is one method of introducing geometry and instilling a value of geometry to people of all ages. Examining polygons with hands-on learning activities and real-world examples provides students with opportunities to investigate the characteristics and properties among polynomial shapes and helps them grasp an understanding of geometry at a higher level.

Development of Polygons

Planar polygons have been important since ancient times. Up until the seventeenth century, polygons that inscribed and circumscribed a circle were used by Archimedes and many others to estimate values of π. In 1796, at the age of 19, Carl Friedrich Gauss constructed a 17-sided polygon using a compass and straight edge. A year earlier, he had described the area of a polygon, which is often referred to as the "Surveyor's formula," although this concept also is attributed to A. L. F. Meister in 1769. The concept of a tiling or tessellation also requires polygons, and these have a long history of representation in art, weaving, architecture, and mathematics. Johannes studied the coverings of a plane with regular polygons, and in 1891, crystallographer E. S. "Yevgraf" Fedorov proved that there are 17 different types of symmetries that can be used to tile the plane. Planar polygons also star as main characters in Edwin Abbott's 1884 novel *Flatland* and the subsequent twenty-first-century movies. In the early twenty-first century, young children investigate the mathematical properties of planar polygons in primary school.

Other types of polygons are also interesting and useful. Non-convex polygons like a star polygon, where line segments connecting pairs of points no longer have to remain inside the polygon, were studied systematically by Thomas Bredwardine in the fourteenth century. Generalized polygons in the twentieth century include complex polygons investigated by Geoffrey Shephard and H. S. M "Donald" Coxeter; Moufang polygons, named after Ruth Moufang; and near polygons. In 1797, Norwegian surveyor Caspar Wessel explored planar and spherical polygons in his theoretical investigation of geodesy. M. C. Escher represented hyperbolic polygons in his tessellated artwork. Some twenty-first-century college geometry texts contain spherical and hyperbolic polygons.

Further Reading

Bass, Laurie E., Basia R. Hall, Art Johnson, and Dorothy F. Wood. *Geometry: Tools for a Changing World*. Upper Saddle River, NJ: Prentice Hall, 1998.

Botsch, Mario, Leif Kobbelt, Mark Pauly, Pierre Alliez, and Bruno Levy. *Polygon Mesh Processing*. Natick, MA: A K Peters, 2010.

Cohen, Marina. *Polygons*. New York: Crabtree Publishing, 2010.

Fowler, Orson. *The Octagon House: A Home for All*. New York: Dover Publications, 1973.

Guttmann, A. J. *Polygons, Polyominoes and Polycubes*. Berlin: Springer, 2009.

Harrell, Marvin E., and Linda S. Fosnaugh. "Allium to Zircon: Mathematics." *Mathematics Teaching in the Middle School* 2, no. 6 (1997).

Icon Group International. *Polygons: Webster's Timeline History, 260 B.C.–2007*. San Diego, CA: ICON Group International, 2009.

Kolpas, Sidney J. *The Pythagorean Theorem: Eight Classic Proofs*. Palo Alto, CA: Dale Seymour, 1992.

Serra, Michael. *Discovering Geometry an Investigative Approach*. Emeryville, CA: Key Curriculum, 2008.

van Maldeghem, Hendrik. *Generalized Polygons*. Basel, Switzerland: Birkhäuser, 1998.
Vogel, Stephen. "How the Pentagon Got Its Shape." May 2007. http://www.washingtonpost.com/wp-dyn/content/article/2007/05/23/AR2007052301296_4.html.
Wells, Katrena. "Hexagons in Nature—6-Sided Shapes Made Fun: Teach Practical Application of Natural Beauty of the Hexagon." http://primaryschool.suite101.com/article.cfm/hexagons-in-nature--6-sided-shapes-made-fun.

Linda Reichwein Zientek
Beth Cory

Polyhedra

Category: History and Development of Curricular Concepts.
Field of Study: Communication; Connections; Geometry.
Summary: Regular solid shapes play important roles in nature and geometry.

People frequently encounter objects in polyhedral shapes, such as buildings that have cubic or prismatic shapes and geodesic domes or dice that are shaped like polyhedra. This prevalence is partly because of their aesthetic appeal and partly because of their practical properties. Polyhedra also appear in nature; many crystals have the shapes of regular solids, particularly of tetrahedron, cube, and octahedron, and virus capsids can be icosahedral. Furthermore, carbon atoms can form a type of molecule known as "fullerenes," which are in the form of a triangulated truncated icosahedron. A polyhedron is a solid in space with polygonal faces that are joined along their edges. If the faces consist of regular polygons, then it is called a "regular polyhedron." A polyhedron is convex if the line segment joining any two points lies on or inside it. Regular convex polyhedra are particularly important for their aesthetic value, symmetry, and simplicity. There are only five of them: the tetrahedron, cube or hexahedron, octahedron, dodecahedron, and icosahedron. Beginning in primary school, students investigate and classify geometric shapes, including polyhedra. In middle school and high school, students explore area and volume measurements as well as transformations and cross-sections.

History

Some of the earliest known polyhedra are the Egyptian pyramids. The five regular solids appear as decorations on Scottish Neolithic carved stone balls, which date to 2000 B.C.E. There are also examples of cuboctahedra worn by east-African women around the ankle and a variety of polyhedral earrings in medieval Europe. The Greeks are thought to have first studied the mathematical properties of regular solids, particularly the Platonic solids, named for Plato. The last book of Euclid of Alexandria's *Elements* is devoted to the study of the properties of these solids, including detailed descriptions of their construction. The book is based on the work of Theaetetus of Athens. There is some evidence that Hippasus of Metapontum may have been the first to describe the dodecahedron. Hypsicles of Alexandria inscribed regular polyhedra in a sphere in his treatise. The Platonic solids also represented physical aspects: Earth was associated with the cube, air with the octahedron, water with the icosahedron, fire with the tetrahedron, and the dodecahedron with the universe. Plato noted: "So their combinations with themselves and with each other give rise to endless complexities, which anyone who is to give a likely account of reality must survey."

The Kepler–Poinsot polyhedra are named for the 1619 work of Johannes Kepler and the 1809 work of Louis Poinsot. They constructed four regular "stellated" polyhedra. These new solids were obtained by extending the faces. In the twentieth century, Donald Coxeter classified and studied the stellation process and described many stellated polyhedra.

Figure 1. Five Platonic solids.

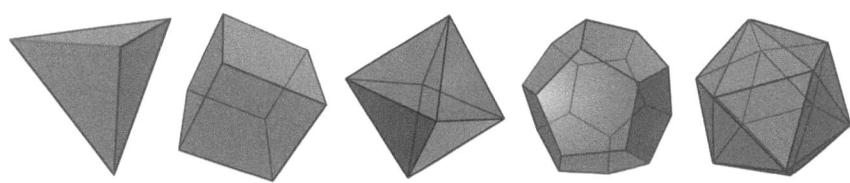

Properties

One common classroom investigation that relates to polyhedra is the Euler characteristic χ, named for Leonhard Euler. It is an equation that combines the number of vertices (V), edges (E), and faces (F) of a polyhedron as $\gamma = V - E + F$. All convex polyhedra have the same Euler characteristic: 2. René Descartes discovered the polyhedral formula in 1635, and Euler discovered it in 1752. In the nineteenth century, Ludwig Schläfli generalized the formula to polytopes and Henri Poincaré proved the result.

The shape of a polyhedron lends itself to a very convenient symbolic or combinatorial description, called the "Schläfli symbol" of the polyhedron. Let $\{n, p\}$ represent a regular polygon with n-gon faces, p of them meeting at each vertex. For example $\{4, 3\}$ would represent a cube because three squares meet at each vertex. This symbolic representation is particularly useful if one would like to express various quantities like the dihedral angle, angular deficiency, radii of inscribed and circumscribed spheres, and surface area. For instance, the surface area of a Platonic solid $\{n, p\}$ can be expressed by

$$S = nF\left(\frac{a^2}{2}\right)\cot\left(\frac{\pi}{n}\right)$$

where F is the number of faces and a is the side length.

Mathematically, polyhedra are very appealing for their fine properties such as duality, symmetry, and versatile constructability. The dual of a polyhedron is constructed by taking the vertices of the dual to be the centers of the faces of the original figure by interchanging faces and vertices. For instance, the dodecahedron and the icosahedron are duals. Many polyhedra are highly symmetrical, and in the nineteenth century, Felix Klein investigated them. The groups of symmetries are algebraic structures consisting of reflections and rotations. One can also generate new polyhedra from old by truncating the vertices of polyhedra, a process known and studied since antiquity. Some of the truncated polyhedra are also known as the "Archimedean solids," named for Archimedes of Alexandria, whose faces consist of two or more types of regular polygons.

There are 13 Archimedean solids, and there are 53 other semiregular, non-convex polyhedra, which are non-Archimedean. The collection of all Platonic, Kepler–Poinsot, Archimedean, and semiregular, non-convex polyhedra together with prisms form the family of polyhedra called "uniform polyhedra."

Non-Euclidean polyhedra took on a prominent role in some theories of a spherical dodecahedral universe at the beginning of the twenty-first century. There are also non-Euclidean polyhedra with no flat equivalents. For instance, a spherical hosohedron with Schläfli symbol $\{2, n\}$ is shaped like a segmented orange or beach ball with lune faces. The name "hosohedron" is attributed to Coxeter.

There have been many artistic and physical models of polyhedra in mathematics classrooms. With the advent of perspective, polyhedra were easier to draw and mathematicians and artists designed and collected polyhedral models. Albrecht Dürer introduced polyhedral nets in his 1525 book. Students continue to use nets to build models. In 1966, Magnus Wenninger published a work on polyhedral models for the classroom through the National Council of Teachers of Mathematics. Wenninger noted that the popularity of the book reflected the continued interest in polyhedra. In the twenty-first century, origami polyhedra have also become important in mathematics and computer science classrooms and research.

Further Reading

Artmann, Benno. "Symmetry Through the Ages: Highlights From the History of Regular Polyhedra." In *Eves' Circles*. Edited by Joby Anthony. Washington, DC: Mathematical Association of America, 1994.

Cromwell, Peter. *Polyhedra*. New York: Cambridge University Press, 1997.

Demaine, Erik, and Joseph O'Rourke. *Geometric Folding Algorithms: Linkages, Origami, Polyhedra*. New York: Cambridge University Press, 2007.

Gabriel, Francois. *Beyond the Cube: The Architecture of Space Frames and Polyhedra*. Hoboken, NJ: Wiley, 1997.

Dogan Comez
Sarah J. Greenwald
Jill E. Thomley

Polynomials

Category: History and Development of Curricular Concepts.
Fields of Study: Algebra; Communication; Connections.
Summary: Polynomial functions have long been studied by mathematicians and have interesting and important applications.

Polynomials have a broad array of theoretical and real-world applications and are widely used by mathematicians, scientists, and engineers to mathematically model data and explore many mathematical and scientific concepts. Technologies that transmit electronic signals, ranging from deep space probes communicating with Earth, to home DVD players, commonly use polynomial error-correcting codes, like the Reed–Solomon codes, named for mathematicians Irving Reed and Gustave Solomon. Cryptographic algorithms that help ensure secure data transmission also rely on polynomials to represent and manipulate data. Calculators may use approximations called "Taylor polynomials," named for mathematician Brook Taylor, for functions like square roots. Civil engineers model and estimate properties, such as volume for lakes and other irregular natural features, with polynomials. Orthogonal polynomials provide the foundation for many multivariate statistical procedures. In twenty-first-century classrooms, polynomials are typically part of advanced middle school or high school curriculums, though linear functions and comparisons of linearity versus nonlinearity are common in middle school, and some of the basic concepts of functions are introduced in the elementary grades.

Early in their mathematical studies, students learn that the graph of the squaring function is a parabola, and that the plot of $y = p(x) = x^2$ is shown in Figure 1, which is the first natural function to consider beyond ones that generate straight lines.

There is an entire family of functions like the squaring function, the cubing function, the fourth power function, and more. If indexed, one could call

the squaring function $p_2(x) = x^2$,

the cubing function $p_3(x) = x^3$,

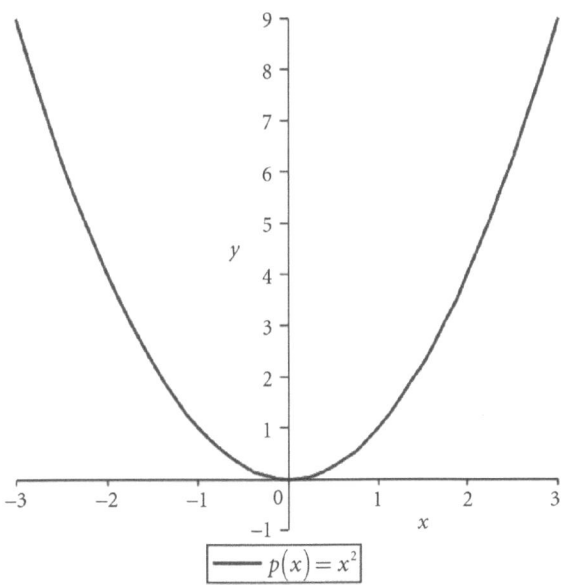

Figure 1.

the fourth power function $p_4(x) = x^4$,

and, in general, the nth power function $p_n(x) = x^n$.

The family of power functions also includes the zero power function $p_0(x) = 1$ and the first $p_1(x) = x^1$. These power functions are the building blocks of "polynomial functions," functions that are made from taking sums and constant multiples of power functions. As such, these functions are especially simple because their formulas only involve addition and multiplication. The first understanding of these power functions is generally credited to Abu Bekr ibn Muhammad ibn al-Husayn Al-Karaji, who lived c. 1000 C.E. in what is now Iraq. In particular, he made advances in the use of variables and humankind's ability to think of arithmetic operations on "placeholders," instead of simply on individual numbers.

Finding the Zeros

Consider this example, $p(x) = x^3 - 2x^2 - 4x + 8$: this function is obtained by taking the cubing function, subtracting twice the squaring function, subtracting 4 times the first power function, and finally adding 8. Regardless of the power functions chosen and the constants multiply by, a polynomial is built. That is, polynomials are functions that have the form

$$p(x) = a_n x^n + \cdots + a_2 x^2 + a_1 x + a_0$$

where $a_0, a_1, \ldots, a_n$ are real numbers. Provided that a_n is not zero, it is stated that p is a degree n polynomial; the degree represents the highest power of x that is present. Much of the modern notational perspective on these functions is due to the work of René Descartes, who in the early 1600s did important work that popularized not only the notation above using subscripts and superscripts but also offered a visual perspective on polynomial functions through their graphs.

Going back to the first polynomial example, $p(x) = x^3 - 2x^2 - 4x + 8$, one can rewrite this sum of multiples of power functions in the formula as a product of even simpler functions. Specifically, it is possible to show that

$$p(x) = x^3 - 2x^2 - 4x + 8 = (x+2)(x-2)(x-2).$$

One can easily observe that $p(-2) = 0$ and $p(2) = 0$. Mathematicians call -2 and 2 the *zeros* or *roots* of $p(x)$; since the $(x-2)$ factor, which leads to the zero 2, appears twice, mathematicians say that "2 is a double root" or "2 is a zero of multiplicity two." The graph of the polynomial in Figure 2 is also enlightening as it shows that the zeros of the function lie where the function crosses or touches the horizontal axis:

Figure 2.

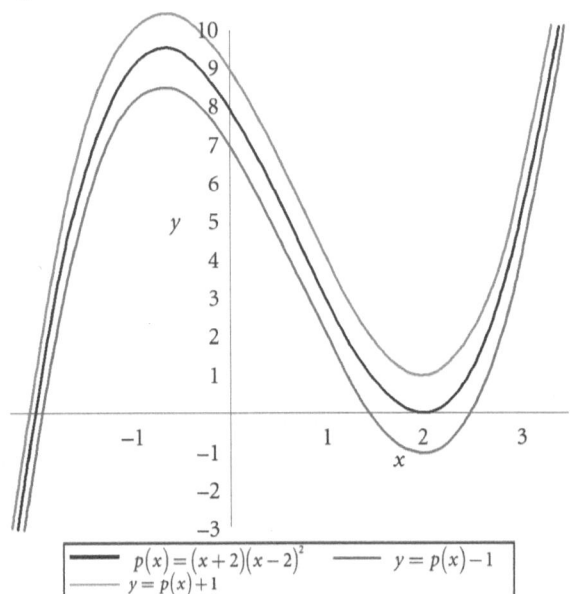

If one shifts the graph of the degree 3 polynomial $p(x)$ (in black) slightly up, the new graph (top line in light gray) will have just one real zero, while if one shifts the graph slightly down, the new function (bottom line in medium gray) will have three distinct real zeros. This illustration demonstrates an important fact about degree 3 polynomials: every degree 3 polynomial has 1, 2, or 3 distinct real zeros. Indeed, the Fundamental Theorem of Algebra, which was proved in its earliest form in 1799 by the great mathematician Carl Friedrich Gauss, states that every polynomial of degree n has at most n distinct real zeros.

If one is willing to permit zeros to be complex numbers and count zeros by their multiplicity, a much stronger version of the Fundamental Theorem of Algebra (which was also known to Gauss) can be proved: every polynomial of degree n has exactly n zeros, provided one counts them according to their multiplicity and allows zeros to be complex. The Fundamental Theorem of Algebra asserts only that n roots of a polynomial function of degree n exist; it does not tell what those roots are.

Quadratic, Cubic, and Quartic Formulas

The search for the zeros of polynomial functions attracted many great minds. The quadratic formula, which calculates the zeros of any degree 2 polynomial, was understood in certain forms by Babylonian mathematicians as early as 2000 B.C.E. The quadratic formula asserts that in order for $ax^2 + bx + c = 0$, it must be the case that

$$x = \frac{-b \pm \sqrt{b^2 - 4ac}}{2a}.$$

For cubic equations and their roots—finding where a polynomial of degree 3 is zero—it took another 3500 years for mathematicians to fully understand the situation. Following contributions from ancient Greeks, Indians, and Babylonians, as well as Persians in the eleventh and twelfth centuries, a group of Italian mathematicians in the 1500s (Scipione del Ferro, Niccolo Tartaglia, and Gerolamo Cardano) proved that there is a cubic formula. In other words, based on the coefficients of a degree 3 polynomial, there is a very complicated formula involving cube roots that calculates the locations of the polynomial's zeros.

Mathematicians were able to take these discoveries a step further. Near the mid-1500s, Ludovico Ferrari found a way to solve quartic equations. This quartic formula is incredibly complicated and represents a major feat in the understanding of polynomial functions. Interestingly, these general formulas cease to exist beyond polynomials of degree 4. In 1824, Neils Abel and Paolo Ruffini published a theorem, based on the work of Evariste Galois, proving that there was no general formula for the roots of a degree 5 polynomial or higher. This latter work on polynomials ended up founding an entire new branch of mathematics called *modern algebra*. Sometimes in mathematics, the quest to solve one problem leads to a whole host of other interesting problems or even a new collection of coherent ideas.

Applications

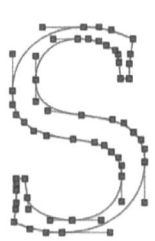

Polynomial functions demonstrate all sorts of interesting patterns and properties and have long been studied because they are interesting in their own right. But even more than this, polynomials play important roles in other areas of mathematics and in applications. For example, polynomial functions spawned the subject of modern algebra, and key ideas in modern algebra are used in the field of public key cryptography—the science of keeping important information private in such essential settings as Internet commerce.

A more direct application of polynomial functions comes in the design of fonts that appear on computer screens. So-called Bezier curves, named for mathematician Pierre Bezier, are degree 3 polynomial functions that can be easily spliced together to form elegant shapes. For instance, at right is the letter S in the Palatino font.

Each piece of the S—the portion of the curve between consecutive squares that represent points on the curve—consists of a degree 3 parametric polynomial. There is deep and elegant mathematics behind why Bezier curves work so well and why they are particularly suited to computer graphics. This is just one example of how substantial ideas and applications in mathematics often emerge from simple beginnings.

Polynomial functions are the simplest of all functions, can be used to approximate more complicated functions that are not polynomials, and often emerge in important applications. They are indeed some of the key building blocks of mathematics.

Further Reading

Barbeau, Edward. *Polynomials*. New York: Springer, 1989.

Kalman, Dan. *Polynomia and Related Realms*. Washington, DC: Mathematical Association of America, 2009.

Kushilevitz, Eyal. *Some Applications of Polynomials for the Design of Cryptographic Protocols*. Berlin: Springer-Verlag, 2002.

Strogatz, Steven. "Power Tools." *New York Times*. http://opinionator.blogs.nytimes.com/2010/03/28/power-tools/.

Matt Boelkins

Probability

Category: History and Development of Curricular Concepts.
Fields of Study: Communication; Connections; Data Analysis and Probability.
Summary: Humans have implicitly understood concepts of probability and randomness since antiquity, but these concepts have been more formally studied since the seventeenth century.

Throughout history, humans have used many methods to try to predict the future. Some believed that the future was already laid out for them by a divine power or fate, while others seem to have believed that the future was uncertain. There are still debates on the extent to which people were able to speculate on the future prior to the development of statistics in the seventeenth and eighteenth centuries. Some assert that such speculations were impossible, yet other historical evidence suggests that at least some people must have been able to perceive the world in terms of risks or chances, even if it was not in quite the same way as later mathematicians and statisticians. The Greek philosopher Aristotle proposed that events could be divided into three groups: deterministic or certain events, chance or probable events, and unknowable events. The idea of "randomness" is often used to indicate completely unknowable events that cannot be predicted. In mathematics, the long-term outcomes of random systems are, in fact, "knowable" or describable using various rules of probability. Probability distributions, expressed as tables, graphs, or functions, show the relationship between all possible outcomes of some experiment or process, like rolling a die, and the chance that those outcomes will happen. For example, lotteries state the chances of winning various prizes, and people seeking medical treatment might be told the odds of success. Random processes and probability can run counter to human intuition and the way in which human brains perceive and organize information, which is perhaps another reason that quantifying ideas of probability is still an ongoing endeavor. Students are often introduced to probability concepts in the earliest elementary grades, such as basic binary classifications of outcomes as "likely" or "unlikely" and the notion of probabilities as experimental frequencies. More formal axioms of probability may be introduced in the middle grades. Probability theory and probability-based mathematical statistics are typically studied in college, though they may be included in advanced high school classes. Some elements of probability theory and applications are also taught in other academic disciplines, like business, genetics, and quantum mechanics.

Early History

Archaeological evidence, such as astragalus bones found at ancient sites, suggests that games of chance have been around for several millennia or longer. Egyptian tomb paintings show astragali being used for games like Hounds and Jackals, much like the way twenty-first-century game players use dice. The ideas of randomness that underlie probability were often closely tied to philosophy and religion. Many ancient cultures embraced the notion of a deterministic fate. The Greek pantheon was among those that included deities associated with determinism, literally known as the Fates. The popular goddess Fortuna in the Roman pantheon suggests a recognition of the role of chance in the world. Jainism is an Indian religion with ancient roots, whose organized form appears to have originated sometime between about the ninth and sixth centuries B.C.E. The Jainist logic system known as *syadvada* includes concepts related to probability; its sanskrit root word *syat* translates variously as "may be" or "is possible." Probability is also a component of the body of Talmudic scholarship; for example, the notion of casting lots, used in some temple functions. Babylonians had a type of insurance to protect against the risk of loss for sea voyages, called "bottomry," as did the Romans and Venetians.

Origins of Study in the Seventeenth Century

Given the near omnipresence of probability in the ancient world, it seems reasonable to think that there were some efforts to estimate or calculate probabilities, at least on a case-specific basis; for example, those who issued maritime insurance would have assigned some type of monetary values for cost and payoff. There is relatively little evidence of broad mathematical research on probability before about the fifteenth century, though some analyses for specific cases survive. For example, a Latin poem by an unknown author called "De Ventula" describes all the ways that three dice can fall. Mathematician and friar Luca Paccioli wrote *Summa de arithmetica, geometria, proportioni e proportionalita* in 1494, which contains some discussion of probability. A few other works address dice rolls and related ideas. Historians tend to agree that the systematic mathematical study of probability as it is now known originated in the seventeenth century. At the time, considerable tensions still existed between the philosophies of religion, science, determinism, and randomness. Determinists asserted that the universe was the perfect work of a divine creator, ruled by mathematical functions waiting to be discovered, and that any apparent randomness was because of faults in human perception. Many emerging scientific theories, like the heliocentric model of the universe advocated by mathematician and astronomer Nicolaus Copernicus, challenged this view by explicitly exploring and quantifying variation and deviations in observations. Astronomy and other sciences, along with the rise of combinatorial algebra and calculus, would ultimately prove to be very influential in the development of probability theory. Changes in business practices also challenged notions of risk, requiring new methods by which likelihood and payoffs could be determined. Harkening back to ancient human activities, however, the most popular story for the origin of probability theory concerns gambling questions posed to mathematician Blaise Pascal by Antoine Gombaud, Chevalier de Méré.

In 1654, the Chevalier de Méré presented two problems. One concerned a game where a pair of six-sided dice was thrown 24 times, betting that at least one pair of sixes would occur. Méré's attempts at calculation contradicted the conventional wisdom of the time and purportedly led him to lose as great deal of money. The second problem, now called the Problem of Points or

Problem of Stakes, concerned fair division for a pot of money for a prematurely terminated game between equally skilled players where the winner of a completed game would normally take the whole pot. Spurred by de Méré's queries, Pascal and Pierre Fermat exchanged a series of letters in which they formulated the fundamental principles of general probability theory.

At the time of its development, Pascal and Fermat's burgeoning theory was commonly referred to as "the doctrine of chances." Inspired by their work, mathematician and astronomer Christian Huygens published *De Ratiociniis in Ludo Aleae* in 1657, which discussed probability issues for gambling problems. Jakob (also known as James) Bernoulli explored probability theory beyond gambling into areas like demography, insurance, and meteorology and he composed an extensive commentary on Huygen's book. One of his most significant contributions was the Law of Large Numbers for the binomial distribution, which stated that observed relative frequencies of events become more stable, approaching the true value, as the number of observations increases. Prior definitions based on gambling games tended to assume that all outcomes were equally likely, which was generally true for games with inherent symmetry like throwing dice. This extension allowed for empirical inference of unequal chances for many real-world applications. Bernoulli also wrote *Ars Conjectandi*. Influenced by this work, mathematician Abraham de Moivre derived approximations to the binomial probability distribution, including what many consider to be the first occurrence of the normal probability distribution, and his *The Doctrine of Chances* was the primary probability textbook for many years.

Objective and Subjective Approaches

Historically and philosophically, many people have asserted that to be objective, science must be based on empirical observations rather than subjective opinion. Estimating probabilities through direct observations is usually called the "frequentist approach." The method of inverse or inductive probability, which allows for subjective input into the estimation of probabilities, is traced back to the posthumously published work of eighteenth-century minister and mathematician Thomas Bayes. Conditional probabilities had already been explored by de Moivre, providing the basis for what is known as "Bayes theorem" (or "Bayes rule"). In Bayes's inductive framework, there is some prob-

The cube design of dice allows for each of their sides to have an equal probability of being rolled. (Photos.com)

ability that a binary event occurs. A frequentist would make no assumptions about the probability and carry out experiments to attempt to determine the true probability value. Using Bayes's approach, some probability value can be arbitrarily chosen, and then experiments conducted to ascertain the likelihood that the value is in fact the correct one. In later interpretations and applications of the method, the initial value might be chosen according to experience or subjective criteria. His work also produced the Beta probability distribution. Bayes's writings contained no data or examples, though they were extended upon and presented by minister Richard Price. At the time, they were relatively less influential than frequentist works, though Bayesian methods have generated much discussion and saw a great resurgence in the latter twentieth century.

Applications

Like Bernoulli, Pierre de Laplace extended probability to many scientific and practical problems, and his probability work led to research in other mathematical areas such as difference equations, generating functions, characteristic functions, asymptotic expansions of integrals, and what are called "Laplace transforms." Some call his 1812 book, *Théorie Analytique des Probabilités*, the single most influential work in the history of probability. The Central Limit Theorem, named for George Pólya's 1920 work and sometimes called the DeMoivre–Laplace theorem, was critical to the development of statistical methods and partly validated the common practice at the time (still used in the twenty-first century) of calculating averages or arithmetic means of observations to estimate location parameters. Error estimates were usually assumed to follow some symmetric probability distribution, such as rectangular, quadratic, or double exponential. While they had many useful properties, they were mathematically problematic when it came to deriving the sampling distributions of means for parameter estimation. Laplace's work, which he proved for both direct and inverse paradigms, rectified the problem for large-sample cases and formed the foundation for large sample theory.

Normal Distribution

The normal distribution is among the most central concepts in probability theory and statistics. Many other probability distributions may be approximated by the normal because they converge to the normal as the number of trials or sample sizes approach infinity. Some of these include the binomial and Poisson distributions, the latter named for mathematician Simeon Poisson. The Central Limit Theorem depends on this principle. Mathematician Karl Friedrich Gauss is often credited with "inventing" the normal (or Gaussian) distribution, though others had researched it and Gauss's own notes refer to "the elegant theorem first discovered by Laplace." He can fairly be credited with the derivation of the parameterization of the distribution, which relied in part on inverse probability. Mathematician Robert Adrain, who was apparently unaware of Gauss's work, discussed the validity of the normal distribution for describing measurement errors in 1808. His work was inspired by a real-world surveying problem. However, Gauss tends to be credited over Adrain, perhaps because of his many publications and the overall breadth of his mathematical contributions.

The fact that Laplace and Gauss worked on both direct and inverse probability was unusual from some perspectives, given the philosophical divide between frequentist and Bayesian practitioners even at the start of the twenty-first century. Later, both would gravitate toward frequentist approaches for minimum variance estimation, which is seen by some as a criticism of inverse probability. Other mathematicians, such as Poisson and Antoine Cournot, criticized inverse methods, while Robert Ellis and John Venn proposed defining probability as the limit of the relative frequency in an indefinite series of independent trials—essentially, the frequentist approach. The maximum likelihood estimation method proposed by Ronald Fisher in the early twentieth century was interpreted by some as melding aspects of frequentist and inverse methods, though he adamantly denied the notion, saying, "The theory of inverse probability is founded upon an error, and must be wholly rejected." This may explain the essential absence of inverse or Bayesian probability concepts in the body of early statistical inferential methods, which were heavily influenced by Fisher.

Mathematician and anthropometry pioneer Adolphe Quetelet brought the concept of the normal distribution of error terms into the analysis of social data in the early nineteenth century, while others like Francis Galton advanced the development of the normal distribution in biological and social science applications in the latter half of the same century. Many mathematicians, statisticians, scientists, and others have contributed to the development of probability theories, far too many to exhaustively list, though recognized probability distributions are named for many of them, such as Augustin Cauchy, Ludwig von Mises, Waloddi Weibul, and John Wishart. Pafnuty Chebyshev, considered by many to be a founder of Russian mathematics, proved the important principle of convergence in probability, also called the Weak Law of Large Numbers. Andrei Markov's work on stochastic processes and Markov chains would lead to a broad range of probabilistic modeling techniques and assist with the resurgence of Bayesian methods in the twentieth century.

Some historians have suggested that one difficulty in developing a comprehensive mathematical theory of probability, despite such a long history and so many broad contributions, was difficulty agreeing upon one definition of probability. For example, noted economist John Keynes asserted that probabilities were a subjective value or "degree of rational belief" between complete truth and falsity. In the first half of the twentieth century, mathematician Andrey Kolmogorov outlined the axiomatic approach that formed the basis for much of subsequent mathematical theory and development. Later, Cox's theorem, named for physicist Richard Cox, would assert that any measure of belief is isomorphic to a probability measure under certain assumptions. It is used as a justification for subjectivist interpretations of probability theory, such as Bayesian methods. There are variations or extensions on probability with many applications. Shannon entropy, named for mathematician and information theorist Claude Shannon and drawn in part from thermodynamics, is used in the lossless compression of data. Martingale stochastic (random) processes, introduced by mathematicians such as Paul Lévy, recall the kinds of betting problems that challenged de Méré and inspired the development of probability theory. Chaos theories, investigated by mathematicians including Kolmogorov and Henri Poincaré, sometimes offer alternative explanations for seemingly probabilistic phenomena. Fuzzy logic, derived from mathematician and computer scientist Lotfali Zadeh's fuzzy sets, has been referred to as "probability in disguise" by Zadeh himself. He has proposed that theories of probability in the age of computers should move away from the binary logic of "true" and "false" toward more flexible, perceptual degrees of certainty that more closely match human thinking.

Further Reading

Devlin, Keith. *The Unfinished Game: Pascal, Fermat, and the Seventeenth-Century Letter That Made the World Modern.* New York: Basic Books, 2008.

Gigerenzer, Gerd. *The Empire of Chance: How Probability Changed Science and Everyday Life.* Cambridge, England: Cambridge University Press, 1990.

Hacking, Ian. *The Emergence of Probability: A Philosophical Study of Early Ideas About Probability, Induction and Statistical Inference.* Cambridge, England: Cambridge University Press, 2006.

———. *An Introduction to Probability and Inductive Logic.* Cambridge, England: Cambridge University Press, 2001.

Hald, Anders. *A History of Probability and Statistics and Their Applications Before 1750.* Hoboken, NJ: Wiley-Interscience, 2003.

Sarah J. Greewald
Jill E. Thomley

Proof

Category: History and Development of Curricular Concepts.
Fields of Study: Communication; Connections; Problem Solving; Reasoning and Proof.
Summary: The product of deductive reasoning, the nature of proof has long been fundamental.

Deductive proofs have been an essential part of mathematics for over 2000 years, and some equate proof ability with competence in mathematics. Mathematician David Henderson defines an effective proof as a convincing communication that answers "why." Thus, a proof may connect ideas within a mathematical system or illuminate both the "how" and the "why" underlying the conjecture. Since a proof depends on the accepted standards of the audience or society through some type of peer review, there is also a long history of concerns about the nature of proof. For example, Galileo Galilei and Christoph Clavius debated the legitimacy of pictures, and Leopold Kronecker criticized the use of nonconstructivist methods. In the twentieth and twenty-first centuries, philosophical concerns about proofs continued as mathematicians considered the role of computers or empirical aspects and the implications of Kurt Gödel's groundbreaking work on consistency. While reasoning and proof have long been a part of mathematics curricula, the concepts took on an increased importance in the United States during the era of Sputnik and the space race, when many different types of proofs were emphasized. In the early twenty-first century, proofs remain fundamental in education, beginning in primary school. In pure mathematics, new research depends on proofs. The notion of proof has been clarified by mathematicians in the field of logic, who explore the foundations of proof.

Brief History

Early civilizations developed sophisticated notions of mathematical argumentation, as documented by evidence such as cuneiform tablets, papyri, and mathematical texts from ancient Babylon, China, and Egypt. The idea of a formal deductive proof arose as a distinct part of ancient Greek mathematics. Greek mathematicians studied and generalized mathematical ideas, using proofs to justify their claims. Mathematics historians theorize that the prevalence of debate in Greek society provided a conducive environment for the development of axiomatic argumentation. Euclid's *Elements* became "the" model for using a small set of axioms to deduce a large system of theorems and knowledge, now known as "Euclidean geometry."

Logic

Logical systems become the foundational structures necessary to create a proof. First, mathematicians use logic tools to argue that one mathematical statement follows as a logical consequence from other mathematical statements, and then they use logic tools to establish a formal proof by building a chain of consequent statements from initial assumptions. These logic tools include connectives (negation, conjunction, disjunction, conditional implications, and equivalence), quantifiers, truth statements, tautologies, and inferential structures (such as *modus ponens* and *reductio ad absurdum*).

Direct Proofs

Mathematical proofs can be done in diverse ways, all reflecting different inferential structures. Starting with an initial conjecture (such as $H \rightarrow C$) involving a hypothesis H and a conclusion C, Direct Proofs build logical chains of compound statements, using conditional implications as the links. They are usually written in the traditional two-column format using high school geometry. The following illustrates the use of a Direct Proof, though it is not entirely rigorous.

Conjecture: If a and b are prime numbers greater than 2, then their sum $a+b$ is composite.
Proof:

Statement	Justification
1. a and b are prime numbers > 2	1. Given
2. a and b are odd numbers	2. The only even prime is 2
3. Let $a = 2m + 1$ and $b = 2n + 1$ for $m, n > 1$	3. Definition of an odd number
4. Sum $a + b = (2m+1) + (2n+1)$	4. Substitution
5. Sum $a + b = 2(m+n+1)$	5. Properties of arithmetic operations
6. Sum $a + b$ is even	6. Definition of an even number
7. Sum $a + b > 2$	7. $m + n + 1 > 1$
8. Sum $a + b$ is composite	8. The only even prime is 2

A Proof by Contrapositive is very similar to a Direct Proof, with the difference being the format of the conjecture itself. While a Direct Proof proves the conjecture $H \rightarrow C$, a Proof by Contrapositive uses a Direct Proof to prove the contrapositive, $\sim C \rightarrow \sim H$. In the previous example, a Proof by Contrapositive would prove the statement "If the sum $a+b$ is prime, then a and b are not both prime numbers greater than 2."

Indirect Proofs

In contrast to Direct Proofs, an Indirect Proof assumes the negation of the conclusion $\sim C$ to be true and then uses a Direct Proof to prove the truth of the negation $\sim S$ for some true statement S. By the Law of Logic Contradiction (S and $\sim S$ cannot both be true), which implies that the original conclusion C must be true. The following illustrates the use of an Indirect Proof.

Conjecture: The $\sqrt{2}$ is an irrational number.
Proof:

Statement	Justification
1. Assume $\sqrt{2}$ is a rational number	1. Negation of conclusion
2. $\sqrt{2} = a/b$ where $\gcd(a,b) = 1$ and a,b positive integers	2. Definition of the rationals and greatest common divisor
3. $2 = a^2/b^2$	3. Squaring both sides
4. $2b^2 = a^2$	4. Multiplying both sides by b^2
5. a^2 is even	5. Definition of an even number
6. a is even	6. Squares of odd integers are odd
7. $a = 2m$ for $m > 1$	7. Definition of an even number

8. $2b^2 = (2m)^2 = 4m^2$	8. Substitution
9. $b^2 = 2m^2$	9. Multiplying both sides by 1/2
10. b^2 is even	10. Definition of an even number
11. b is even	11. Squares of odd integers are odd
12. $\gcd(a,b) \geq 2$	12. Definition of gcd
13. Original assumption is false	13. Contradicting assumption $\gcd(a,b) = 1$

Using the idea of infinite descent, this Indirect Proof is considered to be one of the most "beautiful" proofs by the mathematical community. Though not as elegant, it would be possible to prove this same conjecture using a Direct Proof. Also, it is important to note that this Indirect Proof uses some outside knowledge from number theory (such as, squares of odd integers are odd), which would have to be proved prior to its use as justification within the Indirect Proof.

Deduction and induction constantly play important roles in the proof process. For example, in the previous Direct Proof, a considerable number of examples could be systematically examined: $3 + 5 = 8$, $3 + 7 = 10$, $5 + 11 = 16$, $23 + 47 = 70$, and so on. These examples provide inductive evidence that the conjecture is true, nothing more. That is, the cumulative contribution of the examples is only increased confidence that the conjecture is true and that a formal deductive proof is needed. And, a Proof by Exhaustion of All Cases is not possible because the number of pairs of primes to consider is infinite. Nonetheless, a Proof by Induction is often possible in situations involving an infinite number of examples, as illustrated by the following proof.

Conjecture: $1 + 2 + 3 + 4 + \cdots + n = \dfrac{n(n+1)}{2}$.

Proof:
Case $n=1$: Substituting, $1 = \dfrac{1(1+1)}{2} = 1$.

Assume case for k is true, need to show case for $(k + 1)$ is true: Given the assumption

$1 + 2 + 3 + \cdots + k = \dfrac{k(k+1)}{2}$. Then,

$$1 + 2 + 3 + \cdots + k + (k+1) = \dfrac{k(k+1)}{2} + (k+1)$$

$$= (k+1)\left(\dfrac{k}{2} + 1\right)$$

$$= (k+1)\left(\dfrac{k}{2} + \dfrac{2}{2}\right)$$

$$= \dfrac{(k+1)(k+2)}{2}$$

$$= \dfrac{(k+1)\left[(k+1)+1\right]}{2}.$$

For some conjectures, a visual "Behold!" Proof is possible. A common example is this proof of the Pythagorean Theorem.

Conjecture: In any right triangle, the square of the hypotenuse c (side opposite the right angle) equals the sum of the squares of the other two sides a and b: $c^2 = a^2 + b^2$.

Proof: Behold!

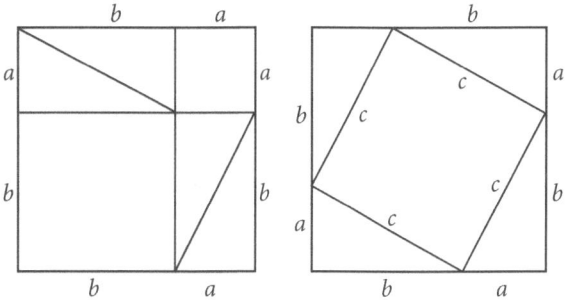

As in most proofs, effort is needed to understand a "Behold!" Proof. In this example, focus on the common areas and rearrangement of the four triangles. The first large square involves two smaller squares (areas a^2 and b^2) and four triangles, while the second large square involves one small square (area c^2) and the same four triangles. Noting that this proof has been traced back to early Chinese mathematics, it is important to add that more than 360 different proofs of the Pythagorean Theorem are known.

Despite their connection to truth, proofs can create mathematical fallacies. Examples include the use of Mathematical Induction to prove that, "All horses are of the same color," the misleading dependence on a geometrical diagram to prove that all triangles are isosceles or even proofs that disguise computational errors, such as the following:

Conjecture: 1=2
Proof:

Statement	Justification
1. Let $n = m > 0$	1. Assumption
2. $n^2 = mn$	2. Multiplying both sides by n
3. $n^2 - m^2 = mn - m^2$	3. Subtracting m^2 from both sides
4. $(n+m)(n-m) = m(n-m)$	4. Factoring both sides
5. $(n+m) = m$	5. Dividing both sides by $(n-m)$
6. $(m+m) = m$	6. Substitution as $n = m$
7. $2m = m$	7. Simplification
8. $2 = 1$	8. Dividing both sides by m

In supporting this obviously wrong conclusion, this proof relies on the reader's literal acceptance of each statement and its justification. That is, the proof seems "true" unless the reader notices that statement five involves division by zero, which is not possible.

When constructing proofs of mathematical conjectures within a system, mathematicians are concerned with many issues related to the logical structure. Is the system "consistent," in that no proven theorem contradicts another? Is the system "valid," in that no mathematical fallacies or false inferences will be created? Is the system based on underlying axioms or initial assumptions that are reasonable? And, is the system "complete," in that every conjecture can be proven either true or false? In the 1930s, logician Kurt Gödel shocked the mathematical world when he proved that a "powerful" mathematical system cannot be both complete and consistent at the same time. For some mathematicians, Gödel's theorems weakened the foundational structure of mathematics, while others felt that it strengthened it. Mathematicians also debate about the role of computers in proofs. In the seventeenth century, Gottfried Leibniz predicted an automatic counting machine that would vastly improve reasoning. In the twentieth century, Herbert Gelernter wrote a program to prove theorems from Euclid's *Elements*, but critics noted the dependence on programmer-supplied rules. Some mathematicians do not accept proofs such as the first proof of the Four-Color Theorem in 1977, which depended on an analysis of many cases by a computer.

Formal proof is a special technique within the realm of mathematics, which is why the public views mathematics as the prime model for establishing truth via argumentation. The idea of proof is invoked in other fields, but with a more limited meaning. For example, in courtrooms, the element of truth is replaced with the phrase "beyond reasonable doubt" given the available evidence. In the sciences, proof is desired but cannot be established by experimental data; at best, the data can support the creation of hypotheses and theories, which will be either further verified or discounted by new experiments and new data.

Further Reading

Cupillari, Antonella. *The Nuts and Bolts of Proofs*. Belmont, CA: Wadsworth Publishing, 1989.

Laczkovich, Miklós. *Conjecture and Proof*. Washington, DC: Mathematical Association of America, 2001.

MacKenzie, Donald. *Mechanizing Proof: Computing, Risk, and Trust*. Cambridge, MA: MIT Press, 2004.

Nelson, Roger. *Proofs Without Words: Exercises in Visual Thinking*. Washington, DC: Mathematical Association of America, 1997.

Polster, Burkard. *Q.E.D.: Beauty in Mathematical Proof*. New York: Walker & Company, 2004.

Solow, Daniel. *How to Read and Do Proofs: An Introduction to Mathematical Thought Processes*. Hoboken, NJ: Wiley, 1982.

Velleman, Daniel. *How to Prove It: A Structured Approach*. Cambridge, England: Cambridge University Press, 1994.

Jerry Johnson
Sarah J. Greewald
Jill E. Thomley

Pythagorean Theorem

Category: History and Development of Curricular Concepts.
Fields of Study: Communication; Connections; Geometry.
Summary: The Pythagorean theorem is a fundamental theorem of mathematics and has numerous applications in number theory and geometry.

The Pythagorean theorem stands as one of the great theorems of mathematics. Ancient peoples appear to have used the Pythagorean theorem to calculate the duration of lunar eclipses or to create right angles in their pyramids or buildings. Archeological evidence suggests that the truth of the result was known in Babylon more than 1000 years before Pythagoras, approximately 1900–1600 B.C.E. Mathematicians and historians continue to debate the early history of the theorem and whether it was discovered independently in such places as Mesopotamia, India, China, and Greece. For instance, some theorize that Pythagoras may have learned the theorem during a visit to India, which in turn may have been influenced by Mesopotamia. The theorem is the culminating proposition of the first book of Euclid's *Elements*. While Euclid (c. 350 B.C.E.) did not mention Pythagoras, later writers such as Cicero and Plutarch referred to it as his discovery. As phrased in the twenty-first century, the theorem states the following:

In any right triangle, the square of the hypotenuse c is equal to the sum of the squares of the legs a and b. That is, $a^2 + b^2 = c^2$.

The theorem has inspired countless generations, and it is useful in a wide variety of contexts and applications, such as in chemistry cell-packing and music.

In Pythagoras's day, humankind had not yet invented algebra. As such, this theorem was not viewed with algebraic perspective but rather in a distinctly geometric way. Visually, as shown in Figure 1, on the right triangle with legs a and b and hypotenuse c, the sum of the areas of the darker gray squares is equal to the area of the lightest gray square.

Proofs

Among the many remarkable features of the Pythagorean theorem, one of the most prominent is that the result admits so many different proofs, including one by former U.S. President James Garfield in 1876. Some

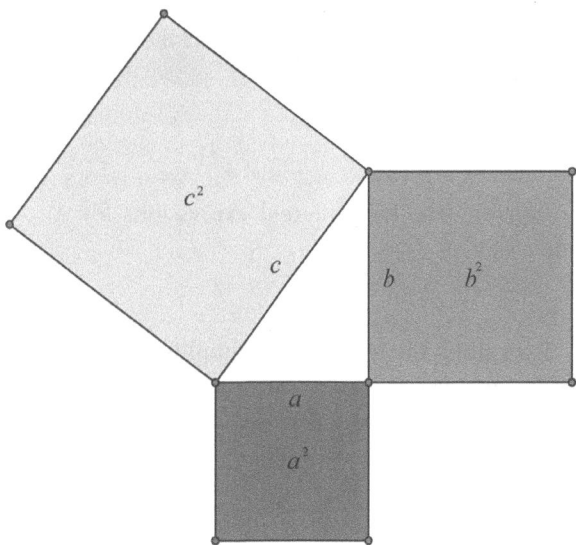

Figure 1.

of the shortest representations of the Pythagorean theorem are geometric figures called "dissections." For example, Indian mathematician Bhaskara's dissection figure was accompanied by the word "Behold." The Chinese also presented dissection figures that are now called "Pythagorean," and some theorize that these may have led to the development of tangram puzzles. Complete Pythagorean proofs based on dissection figures often combine algebra and geometry.

Given a right triangle with legs of length a and b, construct a square of side length $a + b$. Then, along each side, mark a point that lies a units along the side. If consecutive pairs of these points are connected with line segments, four identical (congruent) copies of the original triangle have been constructed inside the large square (see Figure 2).

In addition, these four line segments have generated a quadrilateral (a four-sided polygon) in the interior of the large square. This quadrilateral's sides each have length c, which is the hypotenuse of the given right triangle. Further, a straightforward argument involving angle measurements in the triangles shows that each of the four angles in the interior quadrilateral measures 90 degrees. Hence, the inside quadrilateral is in fact a square.

Consider the area of Figure 2 in two different ways. First, the area A of the entire outside square, which has sides of length $a+b$, must therefore be $A = (a+b)^2$. At the same time, one can view the area of the outside square as having been subdivided into five parts. Four of those pieces are congruent right triangles whose area is each $ab/2$. The fifth part is the interior square, whose area is c^2. Thus, the area A of the outer square also satisfies the relationship that

$$A = \frac{4ab}{2} + c^2.$$

Equating the two different expressions for A, one finds

$$(a+b)^2 = \frac{4ab}{2} + c^2.$$

Expanding the left side and simplifying the right, it follows that $a^2 + 2ab + b^2 = 2ab + c^2$.

Finally, subtracting $2ab$ from both sides, the conclusion of the Pythagorean Theorem follows: $a^2 + b^2 = c^2$.

Applications

Furthermore, the Pythagorean theorem is rightly viewed as one of the most central results in Euclidean geometry. Its statement is equivalent to Euclid's parallel postulate, and therefore is directly tied to the truth of a large number of other key results.

In addition to the geometric ideas the Pythagorean theorem evokes, it generates key new ideas and questions about numbers. For instance, if one takes the legs of a right triangle to each have length 1, then it follows that the hypotenuse c is a number such that $c^2 = 2$. There is no rational number (that is, no ratio of whole numbers) whose square is 2. This situation forced Greek mathematicians to reconsider their original conviction that all numbers were "commensurable": that any possible number must be able to be expressed as the ratio of whole numbers. Remarkably, it took mathematicians another 2000 years to put the so-called real numbers, the set of numbers on which calculus is based, on solid footing.

Another Pythagorean idea that has generated a remarkable amount of mathematics is the notion of a "Pythagorean Triple," which is an ordered triple of whole numbers like (3, 4, 5) that represents a solution to the Pythagorean theorem, since $3^2 + 4^2 = 5^2$. A Babylonian clay tablet, named the "Plimpton 322 Tablet," contains many Pythagorean triples. Some suggest that these were a set of teaching exercises, though historians and mathematicians continue to debate their role. Euclid is credited with the development of a formula that will generate a Pythagorean triple, given any two natural numbers. Indeed, there are even infinitely many "primitive" Pythagorean triples, triples in which

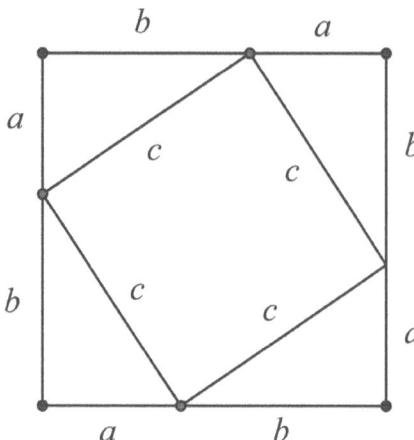

Figure 2.

a, b, and c share no common divisor. Algebraic extensions include investigating solutions to Pythagorean-like equations with other powers, such as $a^3 + b^3 = c^3$. Remarkably, no three positive numbers satisfy such equations; Pierre de Fermat, a French lawyer in the seventeenth century, wrote this note (as translated by historians) in the margins of Diophantus of Alexandria's *Arithmetica*:

> I have discovered a truly marvelous proof that it is impossible to separate a cube into two cubes, or a fourth power into two fourth powers, or in general, any power higher than the second into two like powers. This margin is too narrow to contain it.

No one ever discovered Fermat's proof, yet Fermat's Last Theorem stimulated the development of algebraic number theory in the nineteenth century, and many results in mathematics were shown to be true if Fermat's Last Theorem was true. Andrew Wiles finally proved it to be true near the end of the twentieth century.

There are many other extensions of the Pythagorean theorem. Pappus of Alexandria generalized the theorem to parallelograms. In the 1939 film *The Wizard of Oz*, the Scarecrow recites a version using square roots instead of squares. The Scarecrow's theorem is false in planar geometry, but it can hold in spherical geometry. However, the Pythagorean theorem does not hold on a perfectly round planet. In this case, $a^2 + b^2 > c^2$. Writers for the animated television show *Futurama* named this the Greenwaldian theorem, after mathematician Sarah Greenwald. In the twenty-first century, physicists and mathematicians investigate whether the Pythagorean theorem holds in our universe.

The Pythagorean theorem is also a fundamental idea in several other areas of mathematics and applications. Essentially all of plane trigonometry rests on the Pythagorean Theorem as its starting point, and the modern notion of "orthogonality" in linear algebra is an extension and generalization of the work of Pythagoras. Both trigonometry and orthogonality lead to a wide range of interesting and important applications, including the theory of wavelets and Fourier analysis, mathematics that enables prominent image compression algorithms to help the Internet function.

Its own inherent beauty, the multitude of possible proofs, the rich mathematical ideas it spawns, and the applications that follow all contribute to making the Pythagorean theorem one of the genuine masterpieces in all of mathematics.

Further Reading

MacTutor History of Mathematics Archive. "Pythagoras's Theorem in Babylonian Mathematics." http://www-history.mcs.st-andrews.ac.uk/HistTopics/Babylonian_Pythagoras.html.

Maor, Eli. *The Pythagorean Theorem: A 4000-Year History*. Princeton, NJ: Princeton University Press, 2007.

Posamentie, Alfred. *The Pythagorean Theorem: The Story of Its Power and Beauty*. Amherst, NY: Prometheus Books, 2010.

MATT BOELKINS

Randomness

Category: History and Development of Curricular Concepts.
Fields of Study: Communication; Connections; Data Analysis and Communication.
Summary: While a seemingly simple idea, the concept of randomness has been studied by mathematicians for thousands of years and has many modern applications.

The philosophical concept of determinism supposes that all events that occur in the world can be traced back to a specific precipitating cause and denies the possibility that chance may influence predestined causal paths. Mathematical determinism similarly states that, given initial conditions and a mathematical function or system, there is only one possible outcome no matter how many times the calculation is performed.

Historical Studies of Randomness and Certainty

Many ancient cultures embraced the idea of fate. For example, the Greek pantheon included goddesses known as Fates. At the same time, the existence of ancient gambling games and deities like the Roman goddess Fortuna suggest that these people understood the notion of randomness or chance on some level. Around 300 B.C.E., Aristotle proposed dividing events into three different categories: certain events, which were deterministic; probable events, which were because of chance; and unknowable events.

In the 1600s, the work of mathematicians such as Blaise Pascal and Pierre de Fermat laid some foundations for modern probability theory, which quantifies chance. Abraham de Moivre published *The Doctrine of Chances* in 1718. Around the same time, Daniel Bernoulli investigated randomness in his *Exposition of a New Theory on the Measurement of Risk*. Nonetheless, determinism continued to maintain a prominent place in mathematics and science. Researchers often assumed that seemingly observed randomness in their data was because of measuring error or a lack of complete understanding of the phenomena being observed.

The emergence of fields like statistics and quantum mechanics in the nineteenth century helped drive new work on randomness. Mathematician Émile Borel wrote more than 50 papers on the calculus of probability between 1905 and 1950, emphasizing the diverse ways in which randomness could be applied in the natural and social sciences as well as in mathematics. Applied probabilistic modeling grew very quickly after World War II.

Randomness in Society

In twenty-first-century colloquial speech, the word "random" is often used to mean events that cannot be predicted, similar to Aristotle's unknowable classification. However, probability theory can model the long-term behavior of random or stochastic systems using probability distribution functions, which are essentially sets of possible outcomes having mathematically definable probabilities of occurring. They describe the overall relative frequencies of events or ranges of events, though the specific sequence of individual events cannot be completely determined. Stochastic behavior is observed in many natural systems, such as atmospheric radiation, consumer behavior, the variation of characteristics in biological systems, and the stock market. It is also connected to mathematical concepts like logarithms and the digits of π. Elementary school children discuss some of the basics of randomness when studying data collection methods, like surveys and experiments. Formal mathematical explorations typically begin in high school and continue through college.

Society depends on the use of randomness or the assumption that randomness is involved in a given process. Examples include operating gambling games and lotteries; encrypting coded satellite transmissions; securing credit card data for e-transactions; allocating drugs in experimental trials; sampling people in surveys; establishing insurance rates; creating key patterns for locks; and modeling complex natural phenomena such as weather and the motion of subatomic particles.

Generating Randomness

Generating random numbers, however, is very different from observing random behavior. For example, in 1995, graduate students Ian Goldberg with David Wagner discovered a serious flaw in the system used to generate temporary random security keys in the Netscape Navigator Web browser. Almost every civilization in recorded history has used mechanical systems, such as dice, for generating random numbers and randomness has close ties with gaming and game theory. Physical methods are not generally practical for quickly generating the large sequences of random numbers needed for Monte Carlo simulation and other computational techniques. Flaws in shuffling and physical characteristics, like a worn-down corner on a die, or deliberate human intervention, can also introduce bias. In fact, some people have proven their ability to flip a coin in a predetermined pattern. Motivated by the mathematical unreliability of these physical systems, mathematicians and scientists sought other reliable sources of randomness. Leonard Tippet used census data, believed to be random, to create a table of 40,000 random digits in 1927. Ronald Fisher used the digits of logarithms to generate additional random tables in 1938. In 1955, RAND Corporation published *A Million Random Digits with 100,000 Normal Deviates*, which were generated by an electronic roulette wheel. Random digit tables are still routinely used by researchers who need to perform limited tasks like randomizing subjects to treatment groups in experimental designs as well as in many statistics classes.

The development of computers in the middle of the twentieth century allowed mathematicians, such as John von Neumann, and computer scientists to generate "pseudorandom" numbers. The name comes from the fact that the digits are produced by some type of deterministic mathematical algorithm that will eventually repeat in a cycle, though relatively shorter runs will display characteristics similar to truly random

Casinos monitor roulette wheel performance and rebalance and realign them to keep results random. (Photos.com)

numbers. Using very large numbers, or trigonometric or logarithmic functions, tends to create longer non-repeating sequences. Linear feedback shift registers are frequently used for applications such as signal broadcast and stream cyphers. Linear congruential generators produce numbers that are more likely to be serially correlated, but they are useful in applications like video games, where true randomness is not as critical and many random streams are needed at the same time. Hardware random number generators, built as an alternative to algorithm-driven software generators, are based on input from naturally occurring phenomena like radioactive decay or atmospheric white noise and produce what their creators believe to be truly random numbers.

Randomness Tests

Mathematicians and computer scientists are perpetually working on methods to improve pseudorandom number algorithms and to determine whether observed data are truly random. Randomness can be counterintuitive. For example, the sequences 6, 6, 6, 6, 6, 6 and 2, 6, 1, 5, 5, 4 produced by fair rolls of a six-sided die are equally likely to occur, but most people would say that the first sequence does not "look" random. Irregularity and the absence of obvious patterns are useful ideas, but they are difficult to measure. Distinctions between local and global regularity must also be made, which include the ideas of finite sets and infinite sets. Irenée-Jules Bienaymé proposed a simple test for randomness of observations on a continuously varying quantity in the nineteenth century. Florence Nightingale David published a power function for randomness tests shortly after World War II. Another technique from information theory measures randomness for a given sequence by calculating the shortest Turing machine program that could produce the sequence. The National Institute of Standards and Technology recommends many such tests, including binary matrix rank, discrete Fourier transform, linear complexity, and cumulative or overlapping sums. As of 2010, the digits of π had passed all commonly used randomness tests.

Classical probability theory is not the only way to think about randomness. Claude Shannon's development of information theory in the 1940s resulted in the entropy view of randomness, which is now widely used in many scientific fields. By the latter half of the twentieth century, fuzzy logic and chaos theory also emerged. Fuzzy logic was initially derived from Lotfali Zadeh's work on fuzzy sets and non-binary truth values, while chaos theory dates back to Henri Poincaré's explorations of the three body problem. Bayesian statistics, based on the eighteenth-century work of Thomas Bayes, challenges the frequentist approach by allowing randomness to be conceptualized and quantified as a partial belief, which shares characteristics with fuzzy logic. Spam filtering is one application that relies on Bayesian notions of randomness.

Further Reading

Bennett, Deborah. *Randomness*. Cambridge, MA: Harvard University Press, 1998.
Mlodinow, Leonard. *The Drunkard's Walk: How Randomness Rules Our Lives*. New York: Pantheon Books, 2008.
Random.org. http://www.random.org.

Sarah J. Greenwald
Jill E. Thomley

Sample Surveys

Category: History and Development of Curricular Concepts.
Fields of Study: Communication; Connections; Data Analysis and Probability.
Summary: Mathematicians and statisticians help design sampling methods and techniques to better represent populations and account for biases and missing data.

A survey is a statistical process by which data are collected from a representative sample of some population of interest in order to determine the attitudes, opinions, or other facts about that population. A census is the special case where everyone in the population is surveyed.

For example, the Babylonians are known to have taken a population census around 3800 B.C.E. In one of the first modern surveys, the *Harrisburg Pennsylvanian* newspaper polled city residents about the 1824 presidential election. Polling continued to be largely a local phenomenon until a 1916 national survey by *Literary Digest* magazine, which predicted the winners of several presidential elections despite using highly unscientific survey methods. Their famously incorrect assertion that Alf Landon would beat Franklin Roosevelt in the 1936 election is cited as contributing to the magazine's failure. Journalist and market researcher George Gallup, who correctly predicted Roosevelt's 1936 victory, was a pioneer in statistical sampling in the early twentieth century, though at the time, many considered his ideas quite radical. A post–World War II boom in manufacturing led companies to survey consumers to tailor products to preferences and increase sales. In the twenty-first century, public opinion polls on all aspects of society are pervasive and surveys frequently shape society's opinions and actions in addition to simply measuring them.

Students begin learning how to collect survey data in the primary grades. Researchers in many disciplines also routinely rely on data gathered via surveys. Mathematicians and statisticians work on mathematically valid methods for selecting samples that are random and representative as well as methods to reduce bias in surveys, effectively analyze data, present results that adjust for random error, and account for the effects of missing data. Many of these individuals belong to the Survey Research Methods Section of the American Statistical Association. Leslie Kish, a recipient of the association's prestigious Samuel S. Wilks Award, was especially cited for his worldwide influence on sample survey practice and for being "a humanitarian and true citizen of the world ... [whose] concern for those living in less fortunate circumstances and his use of the statistical profession to help is an inspiration for all statisticians."

History of Surveys

In practice, surveys are collections of questions administered to individuals. Organizations like Gallup (founded as the American Institute of Public Opinion in 1935) specialize in conducting scientifically valid surveys. In the early part of the twentieth century, surveys were mostly conducted door-to-door by trained surveyors, a procedure used by both Gallup and the U.S. Census. Frequently, surveyors used the mail, like in the case of *Literary Digest*. Telephone surveys increased notably in the 1960s, which was attributed in large part to the fact that the costs of in-person research were escalating and trends in non-response suggested that people were growing less willing to answer face-to-face surveys, which diminished their prior advantage over phone surveys. Around 1970, statisticians Warren Mitofsky and Joseph Waksberg developed an efficient method of random digit dialing that revolutionized telephone survey research. However, some major organizations, like Gallup, continued door-to-door surveys into the mid-1980s, at which point they determined that a statistically sufficient proportion of U.S. homes had at least one telephone.

In 2008, Gallup notably expanded its methodology to include cell phones, since an increasing proportion of people no longer use landlines. In the twenty-first century, surveys are increasingly conducted via the Internet, though the U.S. Census still uses a combination of mail and house-to-house surveys. Harris Interactive, which went public in 1999, is a company that specializes in interactive online polls like the Harris Interactive College Football Poll, which ranks the top 25 Bowl Conference Series football teams each week.

Bias

Each survey method has different implications for both response bias and nonresponse bias. It is unclear when mathematicians and pollsters first began to recognize the negative influences of these biases, though adjustments were made in the latter half of the twentieth century. Systematic investigations can perhaps be traced to the mid-twentieth century, coincident with similar concerns in experimental design, like the placebo effect and psychologist Henry Landsberger's naming of the Hawthorne effect. Overall, these biases are problematic because they are non-random and cannot be accounted for by most traditional statistical methods. As a result, they may produce misleading results. Methods to combat these biases are the subject of a great deal of ongoing research and are typically addressed via incentives and proactive planning rather than adjustments after the fact.

Sampling

Randomness is a critical component of survey methodology. Statistical techniques commonly assume that the sample is a random subset of the population. When this is true, the results are more likely to be representative and informative of the population. Though random sampling is the standard in modern scientific polling, early pollsters like Gallup tended to use convenience or quote sampling—taking a sample of whomever was accessible or convenient, sometimes grouped according to other influential variables like political party, gender, or neighborhood. In some cases, this was simply an issue of practicality in terms of time and financial resources. Mathematical statistician Jerzy Neyman is credited with presenting the first developed notion regarding making inferences from random samples drawn from finite populations, what is now called "probability sampling," at a professional conference in 1934. He also contrasted probability sampling with non-random methods. The U.S. Department of Agriculture, in partnership with the statistical laboratory at Iowa State University, began researching probability sampling methods in the late 1930s, as did the U.S. Census Bureau. One of these influential survey researchers was William Cochran, who also helped build many academic statistics programs, including at Harvard. Through the 1940s and beyond, the formal methods of probability sampling and analysis sampling were developed, implemented, and refined in a wide variety of situations.

In the late 1970s and beyond, some researchers' attention turned to more advanced concepts like model-dependent sampling. In probability sampling, the characteristics of the population are wholly inferred from the sample. Model-dependent sampling, in contrast, assumes some probability model for the population beforehand and designs both a sampling and an analysis plan around this model. This method allows the researchers conducting the survey to optimally match the statistical properties of chosen estimators to the population. Statisticians Morris Hansen, William Madow, and Benjamin Tepping discussed many of the principal advantages and limitations of this method in a 1978 presentation and 1983 publication. Morris Hansen was an internationally known expert on survey research, an associate director for research and development at the Census Bureau, and later chairman of the board for polling company Westat, Inc. He also served as president of the American Statistical Association and Institute for Mathematical Statistics.

U.S. Census

Though the U.S. Constitution calls for a count of the population in the decennial census, the U.S. Census Bureau conducts other types of surveys and has been using sampling since 1937. In 1940, the bureau began asking a random sample of people counted in the decennial census extra questions to allow better characterization of population demographics as well as to estimate coverage errors. The ongoing American Community Survey helps determine how billions of federal and state dollars are distributed each year. In the late twentieth century, in large part because of substantial difficulties during the 1990 census, many statisticians proposed completely substituting sampling methods for the decennial counting process or at least substantially increasing the role of sampling. They felt that issues like undercoverage of certain subpopulations could be better addressed with increasingly sophisticated statistical methods. Cost was also considered. They had the support of many cities, states, civil rights groups, and members of Congress. The proposal was opposed by many other politicians and segments of the general population for both political reasons and because of skepticism regarding the sampling process. It ultimately required a ruling by the U.S. Supreme Court, which allowed supplemental sampling for some purposes but required a count to determine congressional apportionment.

Further Reading

Brick, J. Michael, and Clyde Tucker. "Mitofsky–Waksberg: Learning From the Past." *Public Opinion Quarterly* 71, no. 5 (2007). http://poq.oxfordjournals.org/content/71/5/703.full#ref-24.

Hansen, Morris. "Some History and Reminiscences on Survey Sampling." *Statistical Science* 2, no. 2 (1987). http://projecteuclid.org/DPubS/Repository/1.0/Disseminate?view=body&id=pdf_1&handle=euclid.ss/1177013352.

Gareth Hagger-Johnson

Scatterplots

Category: History and Development of Curricular Concepts.
Fields of Study: Communication; Connections; Data Analysis and Probability.
Summary: Scatterplots are useful tools for mathematicians and statisticians to graph and present data.

Human beings are constantly exploring the world around them to discover relationships that can be used to explain past and current events or phenomena and perhaps to predict future occurrences.

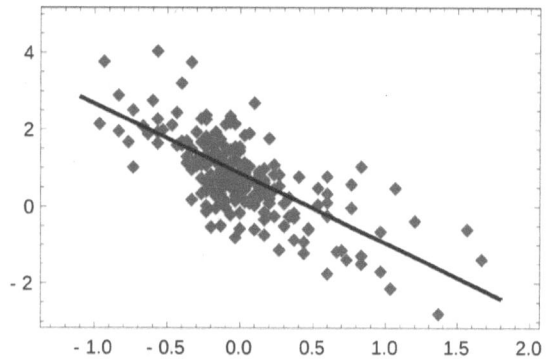

An example of a scatterplot chart.

The colloquial expression "a picture is worth a thousand words" is traced back to many possible historical sources, including French leader and noted student of mathematics Napoleon Bonaparte, who purportedly said, "A good sketch is better than a long speech." In the twenty-first century, graphing is a fundamental first step in any exploratory data analysis, and graphical representations are common in the media. Scatterplots, which most often represent values of paired variables in a Cartesian plane, help data investigators identify relationships, describe patterns and correlation, fit linear and nonlinear functions using techniques like regression analysis, and locate points known as "outliers" that deviate from the predominant pattern. In the primary grades, students often use line graphs, which some consider to be a special case of scatterplots, while scatterplots for data may be explored beginning in the middle grades in both mathematics and science classes.

Early History

Mathematicians and others have long sought alternative methods of representation for researching, presenting, and connecting the mathematical concepts they studied. The Cartesian plane, named for René Descartes, facilitated graphing of algebraic equations and data beginning in the seventeenth century. Historians have traced scatterplots to 1686, though the term "scatter diagram" is attributed to early twentieth-century researchers such as statistician Karl Pearson, and "scatterplot" seems to have first appeared in a 1939 dictionary.

Examples of early pioneers of data graphing include "political arithmetician" Augustus Crome, who studied the relationships between nations' population sizes, land areas, and wealth; mathematician and sociologist Adolphe Quetelet, who conducted studies of body measurements that helped contribute to the measure now known as the Body Mass Index, which relates height and weight; and engineer and political scientist William Playfair, who called himself the "inventor of linear arithmetic," a term he used for graphs. He said: ". . . it gives a simple, accurate, and permanent idea, by giving form and shape to a number of separate ideas, which are otherwise abstract and unconnected." Playfair's eighteenth-century graphical summaries of British trade across various years are perhaps the earliest example of what would now be referred to as "time series plots" (or in some cases "line graphs"), which may be considered a special case of scatterplots.

While Playfair plotted many economic variables as functions of time, the most extensive early use of scatterplots to relate two observed variables is probably the anthropometric and genetic research of Francis Galton, a cousin of scientist Charles Darwin. After studying medicine and mathematics in college, he became interested in the investigation and characterization of variability and deviations in many natural phenomena. He established a laboratory for the measurement and study of human mental and physical traits, focusing on empirical and statistical studies of heredity in the latter half of the nineteenth century. Many of Galton's scatterplots involved graphing parental characteristics on one axis, usually the X, and offspring characteristics on the other. Like scientist Gregor Mendel, some of his initial genetic experiments were conducted on peas; later, he investigated measurements of people. Scatterplots of height appeared in his 1886 publication *Regression Towards Mediocrity in Hereditary Stature*, which is the origination of the name for the statistical technique of regression analysis. The word "mediocrity" in this context was a reference to the mean or average height (not a qualitative judgment) and was used to describe a pattern observed in the data: very short parents tend to have taller children, and very tall parents tend to have shorter children, in both cases closer to the mean.

Recent Developments

Prior to the development of computers and data analytic software, data had to be graphed by hand. In the twenty-first century, computers facilitate many types of scatterplots. In addition to the standard plots of two variables in the Cartesian plane, there are three-dimensional scatterplots that display point clouds to explore the ways in which three variables relate and interact. Symbols used to represent points on a two- or three-dimensional scatterplot may also be coded using different colors or shapes to indicate additional variables and uncover patterns. Matrix plots are square grids of scatterplots for a set of variables that plot all possible pairwise sets, usually arranged such that all of the plots in the same row share the same *Y* variable and all plots in the same column share the same *X* variable. Mathematicians, statisticians, computer scientists, and other types of researchers have explored the theoretical and methodological links between scatterplots and map surfaces for use in applications such as data mining and spatial analysis of geospatial information system (GIS) data.

While they are useful tools for exploration and representation, scatterplots are often subject to misinterpretations. For example, sometimes relationships or correlations shown in scatterplots are mistakenly taken as evidence of cause and effect, which must be inferred from the way in which the data were collected rather than from the strength of the association.

Further Reading

Few, Stephen. *Now You See It: Simple Visualization Techniques for Quantitative Analysis*. Oakland, CA: Analytics Press, 2009.

Friendly, M., and D. Denis. "The Early Origins and Development of the Scatterplot." *Journal of the History of the Behavioral Sciences* 41, no. 2 (2005).

Stigler, Stephen. *The History of Statistics: The Measurement of Uncertainty Before 1900*. Cambridge, MA: Belknap Press of Harvard University Press, 1990.

Gareth Hagger-Johnson

Sequences and Series

Category: History and Development of Curricular Concepts.
Fields of Study: Algebra; Communication; Connections; Number and Operations.
Summary: Sequences and series are important mathematical representations with numerous, interesting applications.

A sequence is a list of objects, called "terms," arranged in a fixed pattern such as 1, 3, 5, 7, 9, . . . or Monday, Tuesday, Wednesday, Thursday, Friday, ... In a series, the terms of a sequence are typically added together. Series have a long history of being used to approximate functions or represent geometric quantities. For example, in the seventeenth century, James Gregory showed how the areas of a circle and hyperbola could be obtained using series. In the early days of calculus, series represented geometric quantities and were manipulated using methods extended from finite procedures. Mathematicians like Niels Abel critiqued the rigor of series and expressed concerns with the foundations of calculus. The theory of series was later made rigorous within the field of analysis. Series are important to many areas in science and engineering. Sequences are explored in the primary and middle grades, while series are introduced in high school.

Famous Sequences

One very famous sequence emerges when considering the reproductive habits of rabbits. Consider two rabbits that are too young to reproduce after their first month of life but can and do reproduce after their second month of life. That pair of rabbits produces another pair after its second month and for each month thereafter. If one assumes that none of the rabbits die and that each pair reproduces in the same manner as the first, the number of pairs of rabbits at the end of each month corresponds to the elements of the sequence 1, 1, 2, 3, 5, 8, 13, 21, 34, ... This sequence is known as the "Fibonacci sequence." It is named after Leonardo de Pisa who was called Fibonnaci, a nickname meaning "son (*filius*) of Bonaccio." He wrote about it in his 1202 book *Liber Abaci*. With the exception of the first two terms, each successive term is found by adding the two terms prior to it. This sequence appears in nature in other situations, including the arrangement of leaves on the stems of certain plants, the fruitlets of a pineapple and the spirals of shells. Some mathematical historians suggest that a Fibonacci-like sequence of integers is also represented in stone balance weights excavated in the 1960s that originated in the eastern Mediterranean during the Late Bronze Age.

Other specific types of sequences have been explored. In 1940, Pavel Aleksandrov introduced a concept called "exact sequences," which found relevance in a wide variety of mathematical fields. In 1954,

Jean-Pierre Serre was awarded a Fields Medal, the most prestigious award in mathematics, in part because of his work on spectral sequences.

Series

A series is often the sum of the terms of a sequence. Series originate as early as the Indian mathematician and astronomer Brahmagupta who gave rules for summing series in his 628 C.E. work *Brahmasphutasiddanta* (*The Opening of the Universe*). The sum of the terms of an arithmetic sequence is called an *arithmetic series*. The arithmetic series 1 + 2 + 3 + 4 + ... + 97 + 98 + 99 + 100 is also a well known one, as it is related to mathematician Carl Friedrich Gauss (1777–1855). At a very young age (around 6 years old), Gauss found the sum of the natural numbers (1, 2, 3, 4, ...) from 1 to 100. That is, the sum given by the series 1 + 2 + 3 + 4 + ... + 97 + 98 + 99 + 100. He was given this task by his teacher to keep him busy while the teacher worked with the other students in the class who were not as mathematically gifted as Gauss. After a relatively short time, Gauss returned to the teacher with the sum 5050. Gauss's method was to pair up the terms of the series. Taking the sum of the first and last term (1 + 100) yields 101. This is the same as the sum of the second and second to last (2 + 99 = 101), the third and third to last (3 + 98 + 101), and so forth. In all, there are 50 such pairs, each of which sums to 101. Thus,

$$1 + 2 + 3 + 4 + \cdots + 97 + 98 + 99 + 100$$
$$= (50)(101) = 5050.$$

Many mathematicians advanced the theory of important series such as power series, trigonometric series, Fourier series, and time series. For example, Nicholas Mercator represented the function $\log(1+x)$ as a series in 1651. Taylor series, named after Brook Taylor, is a representation of a function as an infinite sum of terms calculated from the values of its derivatives at a single point. From the early history of analysis, these power series were important in the study of transcendental functions. Data given as a sequence of data points over time led Wilhelm Lexis to develop time series in 1879.

Applications of Series

Some other series arose in the context of questions related to physics and sparked controversy. The mathematics and physics of a vibrating string and solutions of the wave equation led to trigonometric series. Daniell Bernoulli, Jean Le Rond d'Alembert, Leonhard Euler, and Joseph-Louis Lagrange debated the nature of trigonometric series in the eighteenth century. Joseph Fourier developed Fourier series for the heat equation in the nineteenth century, which was criticized at the time because it contradicted a theorem by Augustin-Louis Cauchy but was explored more rigorously by Johann Dirichlet. An overshoot or ringing in Fourier series was first observed by H. Wilbraham and later explored by Josiah Gibbs. The Gibbs phenomenon has implications in signal processing. The three-body problem, which investigates the behavior and stability of three mutually attracting orbiting bodies in the solar system, was solved by Delaunay in 1860 via representing the longitude, latitude, and parallax of the moon as an infinite series.

However, in 1892, Jules Henri Poincaré showed that these and similar solutions were not in general uniformly convergent, and this criticism created doubt about proofs of the stability of the solar system and eventually led to the formation of the field of deterministic chaos. A prize was offered by King Oscar II of Sweden for a solution to the extension of the three-body problems to n bodies. It has since been proven that no general solution is possible, but the n-body problem was also connected to series in Quidong Wang's 1991 work.

Series were also important as mathematicians searched for efficient ways to represent π and find its digits. Keralese mathematician Madhava of Sangamagramam may have been the first when he used 21 terms of a series and stated π correctly to 11 places. In the 1800s, William Shanks used a series to calculate digits of π in the morning and check them in the evening. He calculated 707 digits of π using this method. However, there was a suspicious lack of the number "7" in the last digits, and it was later found that only the first 527 digits were correct. Johann Lambert used the same series to show in 1761 that π must be irrational—it cannot be expressed as a ratio of whole numbers and has an infinite, non-repeating decimal expansion. Srinivasa Ramanujan found series that converged more rapidly than others, and these efficient series were used as the foundations of computer algorithms.

Binary Series

A very famous series is the binary series that consists of powers of 2: $2^0 + 2^1 + 2^2 + 2^3 + 2^4 + 2^5 + \ldots$. It is theorized that the King of Persia, finding himself very bored, asked that a game be invented for his amusement. The inventor of the game the king found most enjoyable would be given a reward. A servant of the king created the game of chess that was most pleasing to the king. When asked what prize he would like, the servant replied that he wanted grains of rice. The chessboard consists of 64 small squares. As a reward the servant asked for 1 grain of rice for the first square, 2 for the second square, 4 for the third square, 8 for the fourth square and so forth, until all 64 squares had been accounted for. The number of grains of rice requested is the sum $2^0 + 2^1 + 2^2 + 2^3 + 2^4 + 2^5 + \ldots 2^{63}$, and it amounts to 274,877,906,944 tons of rice, which is more rice than has been cultivated on Earth since recorded time. The story goes that the king grew furious at the servant once he knew what was requested. The servant was taking the rice as he received it and distributing it among the poor. At some point, the king indicated that he did not have the rice to pay the servant. The servant indicated that he was content with the amount that he had already received and that it was the king who offered a reward not he who made the initial request. Both parties were pleased.

Applications in Economics

Series appear in many other contexts as well. For example, the future value of an ordinary annuity can be found using a series. An ordinary annuity is an account where an individual makes identical deposits on a regular schedule. The money in the account earns interest that is compounded with the same frequency as the deposits. Suppose an individual deposits $100 every year into an account that earns 6% interest annually. Three years later, the first year's deposit has earned interest over two years, the second account over one year, and the last deposit not at all. The money in the account after three years is given by: $100(1.06)^0 + 100(1.06)^1 + 100(1.06)^2$. The general series can be expressed as a single number

$$A = P\left(\frac{(1+i)^n - 1}{i}\right)$$

where A is the future value of the annuity, P is the payment made at the end of each period, i is the interest rate per period, and n is the number of periods.

Limits

Though infinite sequences consist of infinitely many terms, it may be the case that the sum of the terms of such sequences converges on a given value. Such is the case of the geometric series $.9 + .09 + .009 + .0009 + \ldots$. In this series, the first term is .9 and the common ratio is

$$\frac{1}{10}$$

Applying the formula for the sum of the first n terms of the series yield

$$S_n = \frac{.9\left(1 - \frac{1}{10}^n\right)}{1 - \frac{1}{10}} = \left(1 - \frac{1}{10}^n\right).$$

As the number of terms approaches infinity (as $n \to \infty$), the fraction

$$\left(\frac{1}{10}\right)^n$$

becomes so small that one may consider it zero. Therefore,

$$S_n = \left(1 - \frac{1}{10}^n\right) = 1 - 0 = 1$$

as n grows infinitely large. Since

$$S_n = .9 + .09 + .009 + .0009 + \cdots = .\overline{9}$$

one arrives at the very famous result that: $.\overline{9} = 1$.

Further Reading

Ferraro, Giovanni, and Marco Panza. "Developing Into Series and Returning From Series: A Note on the Foundations of Eighteenth-Century Analysis." *Historia Mathematica* 30, no. 1 (2003).

Klein, Judy. *Statistical Visions in Time: A History of Time Series Analysis, 1662–1938*. Cambridge, England: Cambridge University Press, 1997.

Kline, Morris. *Mathematical Thought From Ancient to Modern Times*. New York: Oxford University Press, 1972.

Laugwitz, Detlef. *Bernhard Riemann, 1826–1866: Turning Points in the Conception of Mathematics.* Boston: Birkhäuser Boston, 2008.

LIDIA GONZALEZ

Similarity

Category: History and Development of Curricular Concepts.
Fields of Study: Communication; Connections; Geometry; Measurement.
Summary: The concept of mathematical similarity has been studied since antiquity.

The concept of "similarity" is universal, playing a particularly large role in the field of geometry. In general, objects may be called "similar" if they share features that look alike, such as shape, color, or value. However, it is a much stronger statement to say that two objects are "mathematically similar." Similarity can be a powerful simplifying assumption in modeling situations. Scaling an object appears in many applications, such as in architecture. Scaling notions can also explain the speed of a hummingbird's heartbeat as compared to a human heart, and why certain insects would collapse under their own weight if they were scaled to a large size. Julian Huxley asserted that the evolutionary struggle to maintain similar surface-to-volume relationships is important in anatomy. Recognizing a similar object is also important. Logician and philosopher Willard Van Orman Quine felt that learning, knowledge, and thought all require similarity so that humans can order objects into categories with similar meaning. Similarity is often connected to triangles in mathematics, starting in grades three through five, but there are many other mathematical situations where it is also useful, such as in the definition of trigonometric functions, in axiomatic arguments, in matrices, in analysis of differential equations, and in fractals.

Early History
Distance calculations contributed to the development of similarity. Thales of Miletus is said to have measured the height of a pyramid using its shadow, but historians are unsure of the method that he used. A method that makes use of similar triangles is attributed to Thales by Plutarch of Chaeronea. In classrooms in the twentieth and twenty-first century, similar experiments are conducted. By measuring the length of the shadow of a tall object, like a pyramid, tree, or building, at the same time as measuring the length of a shadow of a known meter or other stick, a proportion with similar right triangles can be formed. The method assumes that light rays are parallel. In ancient China, instruments such as the L-shaped set-square or gnomon also needed similar triangles. In chapter nine of the *Nine Chapters on the Mathematical Art*, problems were posed and solved using similarity concepts. One of the problems has been translated as

> There is a square town of unknown dimensions. There is a gate in the middle of each side. Twenty paces outside the North Gate is a tree. If one leaves the town by the South Gate, walks 14 paces due South, then walks due West for 1775 paces, the tree will just come into view. What are the dimensions of the town?

Many other mathematicians have worked on a variety of similarity concepts and applications. In Euclid of Alexandria's *Elements*, the various definitions of similarity depend on the figure being examined. Apollonius of Perga explored the similarity of conic sections. During the seventeenth and eighteenth centuries in China, the proportionality of corresponding sides of similar triangles in the plane was quite useful in solving problems in spherical trigonometry. In some twenty-first-century college classrooms, students explore the reason why spherical triangles with shortest distance paths and the same angles must be congruent—there is no concept of similarity on a sphere. Mathematics educators also study the conceptual difficulties in teaching and learning similarity.

Other concepts of similarity arose from mechanics concerns. In his work on the equilibrium of the plane, Archimedes of Alexandria postulated that plane figures that are similar must have similarly placed centers of gravity. Galileo Galilei tried to generalize the notion of geometric similarity to mechanics. Isaac Newton, Hermann von Helmholtz, Joseph Fourier, James Froude, Osborne Reynolds, Lord Rayleigh (John Strutt), and others also worked on variations of similarity in physical situations. Building on their work, and motivated by the lack of a theoretical foundation for flight research, Edgar Buckingham articulated a formal basis for mechanical similarity in 1914. Aside from physical applications, in computer graphics, transformations that preserve similarity can be used to scale mechanical and dynamical behavior in addition to static images.

Further Reading

Fried, Michael. "Similarity and Equality in Greek Mathematics." *For the Learning of Mathematics* 29, no. 1 (2009).

Lodder, Jerry. "Proportionality in Similar Triangles: A Cross-Cultural Comparison." *Convergence*, 2008. http://mathdl.maa.org/mathDL/46/.

Sterrett, Susan. *Wittgenstein Flies A Kite: A Story of Models of Wings and Models of the World*. New York: Pi Press, 2005.

Sarah J. Greenwald
Jill E. Thomley

Squares and Square Roots

Category: History and Development of Curricular Concepts.
Fields of Study: Algebra; Communication; Connections; Geometry.
Summary: Squares and square roots have long challenged mathematicians and have led to various expansions of the number system and developments in number theory.

The square of a number x, denoted x^2, is the number $x \times x$. The inverse operation is called the *square root*: the number x is a square root of y if $y = x^2$, the notation used being $x = \sqrt{y}$. Historically, these operations have been a major source of new problems, ideas, and systems of numbers in the early and modern development of mathematics. Square roots have also appeared in many applications, such as computing the standard deviation of a data set, and have often presented a challenge to scientists and mathematicians in the days before readily available calculating technology. Middle-grade students in the twenty-first century continue to use squares and square roots to simplify computations and solve problems, as do carpenters and engineers.

Definition

Geometrically, the "square" of a number x measures the area of a square whose side has length x. This idea explains the name and is likely the way that ancient civilizations were first confronted with the operation. The Pythagorean theorem is an equality between sums of areas of squares constructed on the sides of a right triangles, namely $a^2 + b^2 = c^2$ if a and b are the two legs and c is the hypotenuse. Applied to the triangle obtained by halving a square of side length one along one of its diagonals, it shows that such a diagonal has length equal to $\sqrt{2}$.

A member of the Pythagorean School sometimes identified as Hippasus of Metapontum (c. fifth century B.C.E.) discovered that this number cannot be expressed as the ratio of two integers—it is irrational. The discovery was a sensation amid the Pythagorean School where it was preached that all numbers were rational and called for an extension of the number system.

In the centuries that followed, extensions of the number system would include all numbers expressible with an infinite number of decimal digits, so that each positive number has a square root (for example, $\sqrt{2} = 1.4142136\ldots$) and negative numbers, which can be multiplied according to the usual associative rules, and the following additional ones governing signs: $-1 \times x = -x$; $(-1) \times (-1) = 1$, which implies that $(-1)^2 = 1$ so that -1 should also be counted as a square root of 1.

More generally, both extensions can be combined to yield the system of real numbers, which are the numbers with sign and infinite decimal expansions. In this system, each square of a number is a positive number (or zero), and each positive number has exactly two square roots, which differ by a sign. For example, 2 has as square roots the numbers $1.4142136\ldots$ and $-1.4142136\ldots$, a fact denoted by the expression $\sqrt{2} = \pm 1.4142136\ldots$.

Computation

Square roots can be computed by hand, by calculator, or by computer (up to the desired numerical approximation) by several methods, including those using sequences, exponentials, logarithms, or continued fractions. Mathematicians in ancient Egypt and Babylonia are some of the first who are thought to have extracted square roots. Early Chinese, Indian, and Greek mathematicians also contributed to this area. According to some historians, the first method to be introduced in Europe was that of Aryabhata the Elder, a Hindu mathematician and astronomer. One of the oldest ones, still at the basis of many currently used algorithms, is the so called Babylonian method (which is also an instance of the modern Newton–Raphson method for solving general equations in one variable). Given a positive number S and choosing an initial "guess" x_0, the method produces a sequence of numbers x_n converging to the square root of S by the rule

$$x_{n+1} = \frac{1}{2}\left(x_n - \frac{S}{x_n}\right).$$

For example, the first approximations to $\sqrt{2}$ starting from $x_0 = 1$ are $x_1 = 15$, $x_2 = 1.416...$, $x_3 = 1.414215...$, $x_4 = 1.4142135623746...$, the last one already having 11 correct decimal digits.

Solving the Quadratic Equation

Square roots are used to solve the general quadratic equation $ax^2 + bx + c = 0$, where a, b, and c are parameters, and a is not zero. The formula, at least partially known to the ancient Greek, Babylonian, Chinese, and Indian mathematicians, is

$$x = \frac{-b \pm \sqrt{b^2 - 4ac}}{2a}$$

provided that the so-called discriminant of the equation, the number $b^2 - 4ac$, is not negative.

Imaginary Numbers

The Italian mathematician Rafael Bombelli, in his book *L'Algebra* written in 1569, proposed the introduction of a new number i, which should denote the square root of -1. Multiplying the number i by real numbers would yield square roots of negative real numbers. The new numbers so obtained are called "imaginary numbers," a name introduced by René Descartes (who meant it to bear a derogatory connotation). A new number system is obtained with the numbers formed by adding a real and an imaginary number; such numbers are called "complex numbers." Complex numbers can be added, multiplied, and divided, and the preceding quadratic formula shows that any quadratic equation has two solutions that are complex numbers; this remains true even if the parameters a, b, c are allowed to be complex number themselves. Actually, a stronger result holds true: Carl Friedrich Gauss (1777–1855) discovered that any equation of the form $a_0 x^d + a_1 x^{d-1} + \cdots + a_{d-1} x + a_d$ has d solutions in complex numbers, an important theorem known as the fundamental theorem of algebra. Partly thanks to this property, complex numbers are of fundamental importance in modern mathematics and in many fields of science and engineering, such as telecommunications.

Implications in Number Theory

Questions regarding squares, square roots, and quadratic forms have played a particularly important role in number theory, often giving rise to the simplest instances of rich theories. Numbers that are squares of integers are called "perfect squares," the first examples being 1, 4, 9, 16, 25, ... Galileo Galilei examined perfect squares in the attempt to understand infinity. Leonardo Fibonacci wrote a number theory book called *Liber Qudratorum*, the book of squares.

The problem of representing integers as sums of perfect squares has also received much attention. Pierre de Fermat (c. 1607–1665) proved that the odd prime numbers that are sums of two perfect squares are exactly those that have remainder 1 when divided by 4, an example being $13 = 2^2 + 3^2$ (whereas, for example, the prime number 7 has no such representation). Joseph Louis Lagrange (1736–1813) proved that every positive integer can be written as the sum of at most four perfect squares (for example, $15 = 9 + 4 + 1 + 1$); three squares suffice only for those numbers which are not of the form $4^k(8m + 7)$, as was later proved by Adrien-Marie Legendre.

In his 1801 masterpiece *Disquisitiones Arithmeticae*, written at the age of 21, Gauss investigated two problems whose generalizations are still major topics of current research. The first one is related to the question of representing integers as the sum of squares and asks for a classification of binary quadratic forms, which are functions of two variables x and y of the shape $f(x, y) = ax^2 + 2bxy + cy^2$, where a, b, and c are integer parameters, in terms of the set of integers they represent—the set of possible values of $f(x, y)$

as x and y range among the integers. The second problem considered by Gauss is the following: given two odd prime numbers p and q, is it possible to write p as the difference of a perfect square and a multiple of q (in symbols $p = n^2 - mq$)? Conversely, is it possible to write q as the difference of a perfect square and a multiple of p? Gauss proved that if at least one of p, q leaves remainder 1 when divided by four, then the two questions have the same answer; and that if p and q both leave remainder 3 when divided by 4, then the answer to the second question is "no" whenever the answer to the first question is "yes" and vice versa.

As a consequence of this result (known as the "quadratic reciprocity law") he was able to give an efficient method for answering the question. In fact, Gauss found not one but eight different proofs of this fact, which is so central in modern number theory that about 200 more proofs were later found.

Further Reading
Conway, John H., and Francis Y. C. Fung. *The Sensual (Quadratic) Form*. Washington, DC: Mathematical Association of America, 1991.
Mazur, Barry. *Imagining Numbers: (Particularly the Square Root of Minus Fifteen)*. New York: Farrar, Straus and Giroux, 2003.
Nahin, Paul J. *An Imaginary Tale: The Story of i*. Princeton NJ: Princeton University Press, 2010.

Daniel Disegni

Surfaces

Category: History and Development of Curricular Concepts.
Field of Study: Communication; Connections; Geometry.
Summary: Surfaces are two-dimensional manifolds, some of which have been studied for their special properties.

Living beings interact with much of the world through surfaces. Humans walk on surfaces and eat and sleep on them. Surfaces like the one-sided Klein bottle, named for Felix Klein, stretch the imagination and are the subject of mathematical investigations. They are often represented using physical models as well as computer models, including sculptures and computer animations. In twenty-first-century classrooms, students investigate a variety of surfaces and their properties, including area and volume. The National Council of Teachers of Mathematics recommends an understanding of the area and volume of rectangular solids for primary school students; of prisms, pyramids, and cylinders for middle grades students; and of cones, spheres, and cylinders for high school students. The parametrization and volume of surfaces is further explored in a multivariable calculus course.

History of Study
Mathematicians have long developed the theory of surfaces, and they continue to investigate their properties. In addition to the plane, polyhedra, such as the surface of a cube or an icosahedron, are among the first surfaces studied by the ancient Greeks in geometry. Their view of surfaces was entirely different from the functional description used in investigations of surfaces in the twenty-first century. The Greeks also had a good knowledge of surfaces of revolution and pyramids. With the introduction of analytic geometry in the seventeenth century, the study of surfaces developed into one of the most studied branches of mathematics. Mathematicians like Carl Friedrich Gauss, Pierre Bonnet, Barnhard Riemann, Gaspard Monge, and their followers firmly established surfaces on a rigorous basis during the eighteenth and nineteenth centuries.

One of the greatest achievements of the theory of surfaces is the Gauss–Bonnet theorem. Versions of the theorem were explored by Gauss in the 1820s and Bonnet and Jacques Binet in the 1840s. The form of the theorem that is standard in undergraduate differential geometry courses is attributed to Walther von Dyck in 1888. Smooth surfaces are defined as those surfaces in which each point has a neighborhood diffeomorphic to some open set in the plane. This added structure allows the use of analytic tools. The parametric functions for a smooth surface define two quadratic differential forms: the first and second fundamental forms, which are local invariants defined as functions of the arc length. The Gaussian curvature of the surface is an isometric invariant; hence, an intrinsic property of a surface, which is known as the Theorema Egregium of Gauss. The Gaussian curvature measures the deviance of the surface from being flat at each point. The parametric equations of a surface determine all six coefficients of its first and second fundamental forms; conversely, the fundamental theorem of surfaces states that given six functions satisfying certain compatibility conditions, then there exists a unique surface (up to its location in space). A geodesic curve on a smooth surface is characterized as a locally minimizing path. In a sense, geodesics are the straight lines of surfaces, which are essential for defining the notions of distance, area, and angle on a surface. Bonnet investigated the geodesic curvature, which measures the deviance of a curve on the surface from being a geodesic. The Gauss–Bonnet theorem states that for an orientable compact surface, the total Gaussian curvature is 2π times the Euler characteristic of the surface, named for Leonhard Euler. A consequence of this theorem is that the sum of the interior angles of a geodesic triangle is greater than, less than, or equal to π, depending on if the Gaussian curvature of the surface is positive, like on a sphere; negative, like on a hyperboloid; or zero, like in the plane.

Types of Surfaces

The classification of surfaces is another topic that is explored in undergraduate geometry or topology classes. In 1890, Felix Klein asked what surfaces locally look like the plane. In Klein's Erlangen Program, a space was understood by its transformations. Heinz Hopf published a rigorous solution in 1925 that arose from groups of isometries acting on the plane without fixed points. The surfaces are the plane; the cylinder; the infinite Möbius band, named for August Möbius; the flat Clifford torus or donut, named for William Clifford; and the flat Klein bottle. Intuitively, surfaces seem to always have two sides; however, the Möbius band and the Klein bottle have only one side. Other surfaces like the projective plane resemble the sphere, and many-holed donuts resemble hyberbolic space. The Euler characteristic is used to classify the topology of a surface.

Some important types of surfaces that are studied intensively in geometry and analysis are minimal surfaces with zero mean curvature, such as catenoids and helicoids; developable surfaces with zero Gaussian curvature, like the plane, the cylinder, the cone, or a tangent surface; and ruled surfaces that can be generated by the motion of a straight line, like the cylinder and the hyperboloid of one sheet. While some of these surfaces date to antiquity, others are more recent. In the eighteenth century, Euler described the catenoid and Jean Meusnier the helicoid. Discoveries in 1835 by Heinrich Scherk and in 1864 by Alfred Enneper included minimal surfaces that are now named for each of them. In the 1840s, Joseph Plateau's experiments indicated that dipping a wire ring into soapy water will create a minimal surface. Jesse Douglas won a Field's Medal in 1936 for his solution to Plateau's problem in minimal surfaces. A minimal surface that originated at the end of the twentieth century is because of Celso Coasta in 1982.

Representations and Investigations

Algebraic geometers investigate algebraic surfaces, such as cubic or quartic surfaces that can be represented by polynomials. These led to rich mathematical investigations in the nineteenth and twentieth centuries. For instance, in 1849, Arthur Cayley and George Salmon showed that there were 27 lines on a smooth cubic surface. Quartic surfaces were of interest in optics, and mathematicians such as Ernst Kummer studied them.

While mathematicians had long built physical models of surfaces, which were typically housed in universities and museums, in the late twentieth and early twenty-first centuries, computer-generated surfaces revolutionized the visualization and construction of surfaces and led to many interesting mathematical questions. For instance, numerous mathematicians and computer scientists have explored the method of subdivision of surfaces, including Tony DeRose, and Jos Stam, who won a Technical Achievement Award in 2005 from the Academy of Motion Picture Arts and Sciences. This method often takes advantage of the similarity between the local structure of a surface and a small piece of a plane. For instance, surfaces may be represented using small flat triangle or quadrilateral mesh representations. These representations are easier to manipulate, but they can still appear smooth to the eye.

Further Reading

Andersson, Lars-Erik, and Neil Stewart. *Introduction to the Mathematics of Subdivision Surfaces*. Philadelphia, PA: Society for Industrial and Applied Mathematics, 2010.

Farmer, David, and Theodore Stanford. *Knots and Surfaces: A Guide to Discovering Mathematics*. Providence, RI: American Mathematical Society, 1996.

Fischer, Gerd. *Mathematical Models: Photograph Volume and Commentary*. Braunschweig, Germany: Friedrick Vieweg, 1986.

Gottlieb, Daniel. "All the Way With Gauss-Bonnet and the Sociology of Mathematics." *American Mathematical Monthly* 104, no. 6 (1996).

Henderson, David W. *Differential Geometry: A Geometric Introduction*. Upper Saddle River, NJ; Prentice Hall, 1998.

Pickover, Clifford. *The Math Book: From Pythagoras to the 57th Dimension: 250 Milestones in the History of Mathematics*. New York: Sterling, 2009.

Dogan Comez
Sarah J. Greenwald
Jill E. Thomley

Trigonometry

Category: History and Development of Curricular Concepts.
Fields of Study: Communication; Connections; Geometry; Measurement.
Summary: Trigonometry is one of the most essential branches of mathematics in engineering and science.

The principal value of trigonometry rests in its numerous and practical applications throughout the world. It has always been viewed as one of the most applied areas of mathematics. When combined with the latest technologies, trigonometry impacts society in immeasurable ways. It is essential in engineering and in all of the sciences. Some of the applications of trigonometry include astronomy, aviation, architecture, engineering, geography, physics, seismology, surveying, oceanography, cartography (mapmaking), navigational systems, space sciences, medical imaging, music, and video games. Clearly, the applications and influences of trigonometry are fully embedded within contemporary society. To appreciate modern applications of trigonometry and the contributions of societies throughout the world, it is important to consider its historical evolution.

Origins of Trigonometry

The beginnings of trigonometry date back to prehistoric cultures and mirror the evolution of civilization itself. The word "trigonometry" comes from the Greek word *trigonon* (meaning "triangle") and *metron* (meaning "measurement"). However, trigonometry did not originate as the study of triangles. It was initially viewed as a combination of geometry and astronomy used in studying the movements and locations of celestial bodies in the sky and in time keeping. The foundations of trigonometry originated in prehistoric cultures with their vigilant observations of the night sky. Some of their discoveries are inherent in the designs of Stonehenge and ancient Egyptian monuments. The early civilizations of Egypt, Babylonia, and India (c. 2000 B.C.E.) contributed significantly to the origins of trigonometry. The Egyptians applied the properties of geometry and astronomy in constructing the pyramids, and the Babylonians (c. 1900 B.C.E.) used angles and ratios to keep track of the motions of celestial bodies in the sky. They followed the paths of planets, the lunar and solar eclipses, and gave coordinates to the stars, all of which required familiarity with angular distances measured on the "celestial sphere." While these early civilizations used trigonometric principles for astronomical measurements, in designing monuments, and in time keeping, they did not fully develop the mathematical system that is now known as "trigonometry."

The invention of trigonometry emerged as a defined body of knowledge from work conducted by Greek astronomers around 350 B.C.E. in the city of Alexandria, Egypt, the intellectual center of the ancient world. The Greek astronomer, Hipparchus (190–120 B.C.E.), is frequently called "the Father of Trigonometry. " He used ratios to determine the distances of the Earth from the sun and the moon and was responsible for tabulating the measures of arcs and their corresponding chord lengths for angles within a circle. The trigonometry of ancient Alexandria would now be called "spherical trigonometry" since it was confined primarily to the study of the properties of great circles on spherical bodies.

The earliest recorded work on spherical trigonometry appears in the book, *Sphaerica*, written by Menelaus, a Greek mathematician working in Alexandria (70–140 C.E.). It describes Menelaus' theorem, which associates the ratios of the lengths of intersecting arcs of great circles on a sphere.

Development of Trigonometric Functions

A fundamental difference between ancient Alexandrian trigonometry and modern trigonometry is that the former used arcs and chords in its trigonometric tables instead of the trigonometric functions that are used in the twenty-first century: sine, cosine, and tangent. The oldest surviving table of arcs and chords was created by Ptolemy of Alexandria (90–168 C.E.). However, when the chords of the circle are rotated to a vertical position and radii are inserted to connect the endpoints and midpoint of the chord to a common center, it is possible to translate half-chord lengths to a sine function. Thus, Ptolemy's famous table of arcs and chords was equivalent to a modern table of sines. His most famous mathematical work was the *Almagest*. It included a table of chords for angles for one-half degree to 180 degrees, in increments of half degrees. His chords were accurate in length to five significant digits. In addition, Ptolemy proved (using chords) the formulas for the sum and difference of two angles that are equivalent to the current sine functions of two combined angles. These discoveries were applied to astronomy, time keeping, and in locating the direction of Mecca for the daily five prayers required by followers of Islam.

In the first century B.C.E., trigonometric principles were used primarily in navigation and map-making. Fortunately, Ptolemy recorded all of the geographical knowledge collected by the ancient world in his eight volumes, titled *Geographia*. It included the latitudes and longitudes of 8000 places on Earth and was the world's first atlas, similar to those used in the twenty-first century. It took astronomers nearly 400 years to shift from using tables of angles and chords to a reliance on tables of sines. Indian astronomers of the fifth century gave trigonometry its current interpretation of the sine function, which quickly spread to Arab and Islamic astronomers.

It is important to recognize that the six trigonometric functions that are used in the twenty-first century were developed at the end of the tenth century by Arab and Islamic astronomers. The West learned about Arab and Islamic trigonometry at the beginning of the twelfth century through translations of Arabic and Islamic astronomy handbooks. Indeed, the maps created by the Alexandrian Greeks and the trigonometric functions developed by the Arab and Islamic astronomers were employed during the world explorations of the fifteenth and sixteenth centuries. Christopher Columbus (1451–1506) utilized these materials to guide him to discover the "new world" in 1492. There is no denying the meritorious impacts of trigonometry on exploration and navigation during the fifteenth and sixteenth centuries.

Trigonometry of the Renaissance

Calculations using trigonometric functions to determine the sides of right triangles did not become prevalent until the sixteenth century. German mathematician Bartholomew Pitiscus (1561–1613) is recognized for creating the word "trigonometry," meaning triangle measurement. He chose *Trigometria* for the title of his book, which described the applications of trigonometry in surveying.

During the seventeenth century, numerical calculations in trigonometry were simplified by the invention of logarithms by the Scottish mathematician John Napier (1550–1617). Fifty years following Napier's invention of logarithms, English mathematician Isaac Newton (1642–1727) invented the calculus in which he represented trigonometric functions as infinite series in powers of *x*. Specifically, Newton discovered that

$$\sin(x) = x - \frac{1}{6}x^3 + \frac{1}{120}x^5 - \frac{1}{5040}x^7 + \frac{1}{362880}x^9 - \cdots$$

and

$$\cos(x) = 1 - \frac{1}{2}x^2 + \frac{1}{24}x^4 - \frac{1}{720}x^6 + \frac{1}{40320}x^8 - \cdots.$$

These representations of the sine and cosine functions continue to play important roles in applied and pure mathematics in the twenty-first century.

For most of the history of trigonometry, angles were measured in degrees, which were defined as fractions of the circumference of a circle. This practice was not efficient nor consistent because the radius of the circle was not fixed. The creators of tables of sines often chose a radius convenient for their calculations. Ptolemy used a radius of 60 since his fractions were expressed in 60ths. The Austrian mathematician Georg Rheticus (1514–1574) used a radius of 10^{15}, which permitted him to tabulate the six trigonometric functions with 15-digit accuracy without the use of decimals or fraction manipulation.

Further innovations and interpretations in trigonometry were developed in the eighteenth century by the Swiss mathematician Leonhard Euler (1707–1783). He explained that computations of trigonometric functions would be more efficient if the line lengths were measured in the same unit. Consequently, he chose "1" to be the radius of a circle centered at the origin. Thus, the circle's circumference would be 2π; the arc for 45 degrees would be π/4; the arc for 30 degrees would be π/6; and so on. This system led to the later development of radian measures for angles and arc lengths.

More significantly, Euler discovered that trigonometric functions could be defined in terms of complex numbers (the union of real and imaginary numbers). Specifically, Euler's famous equation states that for any real number x,

$$e^{ix} = \cos(x) + i\sin(x).$$

The relevance of this equation was that trigonometry could then be viewed as only one of the numerous applications of complex numbers. This formula served as a unifying concept for the study of mathematics as a whole.

Contemporary Applications of Trigonometry

Although trigonometry originated in ancient civilizations from studying the movements of astronomical bodies and triangle relationships, its current applications encompass far more. Trigonometry now spans the diverse fields of architecture, engineering, science, music, navigation, medicine, digital imaging, and games of entertainment.

Trigonometry is a perfect partner for architecture and engineering. Contemporary buildings with curved surfaces in glass and steel would be impossible without trigonometry. Although the surfaces are perceived as curved, they are frequently composed of numerous triangles. Furthermore, since the triangle is an ideal shape for evenly distributing the weight of a structure, an understanding of relationships among the parts of triangles is essential in the design and construction of buildings, bridges, and monuments. Specifically, if an engineer knows the lengths of the beams that will be attached to a structure, the angles at which they must be attached can be calculated using trigonometry. Additionally, in the architectural design of an amphitheatre, the engineer's task is to design the structure so that all sounds from the stage are funneled into the audience's ears. Engineers and architects use trigonometry to identify the perfect shape to balance this sound as it reflects off the walls and ceilings.

Since trigonometry facilitates the understanding of space, it has numerous applications in the physical sciences. In optics and statics, trigonometric functions are vital in understanding the behavior of light and sound. It is particularly useful for modeling the periodic processes found in music because of the cyclical and periodic nature of trigonometric functions (sine, cosine, and tangent). Harmonics are determined by the form of their sine waves with respect to their periods and frequencies. The period of a sine wave is the length of the interval of repetition of the sine wave. The frequency of a sine wave is the number of cycles a sine wave goes through in a standard distance or time interval. In music, the frequency is often expressed in units of hertz, (Hz), where 1 Hz means one period per second. For example, the pitch of every note in music is determined by the length of its sine wave (period) and by its frequency. Musical notes with wide sine waves are lower in pitch because they have fewer cycles per second, while notes that have narrow sine waves are higher in pitch and have more cycles per second.

Trigonometry plays a vital role in modern technologies through the process of triangulation. It is used by global positioning systems (GPS), in computer graphics, and in gaming. Specifically, computer generation of complex images is made possible by coloring numerous, microscopic squares (called "pixels") that define the precise location and points on the image. The technique of triangulation is used to make the image highly detailed and clearly focused. In GPS, triangulation is used for object location. Similar imaging technologies have also revolutionized the medical fields through the development of Computed Axial Tomography (CAT) scans, ultrasounds, and magnetic resonance imaging (MRI).

Trigonometry Instruction

It has been shown that trigonometry emerged approximately 4000 years ago through careful observations of the movements of celestial objects in the sky. These observations gave rise to the study of the celestial sphere and relationships among arcs, chord lengths, and central angles of great circles. In the twenty-first century, this form of trigonometry is called "spherical trigonometry." It was not until the end of the tenth century that Arab and Islamic astronomers defined the six trigonometric ratios. It took another 500 years before right triangle trigonometry became prominent in surveying, navigation, and architecture throughout the Western world. Therefore, it is interesting that this evolutionary sequence of the development of trigonometry is not reflected in the order in which it is typically introduced to students.

The current order of instruction is to first introduce the six trigonometric functions as ratios between the sides of right triangles. These concepts are then applied in finding the missing parts of right triangles. Finally, the unit circle and the periodic nature of trigonometric functions are introduced. This instructional order is often supported by the fact that right triangle trigonometry is a natural extension of the study of the Pythagorean theorem and ratios between parts of triangles, both of which are covered in the middle grades. Furthermore, trigonometric ratios are not as difficult to comprehend as periodic functions.

Some educators may argue that instruction should follow the evolutionary sequence of a topic. In support of that argument, students of today's technological world (with microwaves, wifi's, electrocardiograms, and electronic music) gain early familiarity with the periodicity of the sine and cosine functions. They are also more likely to model scientific and social phenomena with trigonometric functions than to apply right triangle trigonometry in finding the missing parts of triangles. Thus, for research purposes in identifying "best practices" for instruction, some educators have considered assessing the effects of teaching trigonometry in the same order that it historically evolved.

In conclusion, trigonometry is more valuable to society today than ever before in recorded history. The foundations of trigonometry emerged about 4000 years ago in the Egyptian and Babylonian civilizations. It developed into a well-defined mathematical discipline in the city of Alexandria, Egypt, circa 350 B.C.E. In the centuries that followed, trigonometry continued to evolve through the contributions and insights of diverse cultures and societies throughout the world. Trigonometry, with its advanced measurement facilities and associated technologies, remains one of the most applicable and practical fields of mathematics, vital to the advancement of the sciences, engineering, and technologies.

Further Reading

Baumgart, John K., ed. *Historical Topics for the Mathematics Classroom*. 2nd ed. Reston, VA: National Council of Teachers of Mathematics, 1989.

Bressoud, David M. "Historical Reflections on Teaching Trigonometry." *Mathematics Teacher* 104, no. 2 (September 2010).

Buchberg, Jerrold T., et al. *The Essential Physics of Medical Imaging*. 2nd ed. Philadelphia: Lippincott Williams & Wilkins, 2002.

Joyce, David. "Applications of Trigonometry" (1997). http://www.clarku.edu/~djoyce/trig/apps.html.

Katz, V., ed. *The Mathematics of Egypt, Mesopotamia, China, India, and Islam*. Princeton, NJ: Princeton University Press, 2007.

Van Brummelen, Glen. *The Mathematics of the Heavens and the Earth: The Early History of Trigonometry*. Princeton, NJ: Princeton University Press, 2009.

Sharon Whitton

Units of Area

Category: Space, Time, and Distance.
Fields of Study: Algebra; Geometry; Measurement; Number and Operations; Representations.
Summary: Numerous units of area have been used throughout history for measuring land.

Specific measurements of land area date back to ancient times to define land ownership (for the purposes of taxation, among other reasons). Some of these measurements are still used in the twenty-first century.

Ancient Units of Measurement

In Mesopotamia, land area was divided into a *bur* (an estate), which covered about 64,800 square meters. The bur, in turn, was divided into *iku* (fields), each of which covered about 3600 square meters. Further measurements and sub-divisions are recorded on surviving land documents. The Egyptians also had their own system based on the *kha-ta* (100,000 square cubits), which in turn was divided into 10 *setat*, which consisted of 10 *kha* (1000 square cubits, or 275.65 square meters).

The Romans had a very specific system of measuring land with the basic measure being an *actus quadratus* (acre), which covered about 1260 square meters. Smaller measurements were described as being a *pes quadratus* (square foot) or *scripulum* (or square perch). These measurements were based on the *pes* (foot) being the basic unit of measurement throughout the Roman Empire, a length that was fixed throughout the Empire.

By contrast, the Greeks used a different system of land measurement by which land was divided into a "plethron"—a variable area of land that consisted of the amount of land a yoke of oxen were able to plough in a single day. As a result, the exact measurment varied from some parts of Greece to other parts (and indeed for different parts of a city), although it was thought to approximate to about four English acres. In rocky and hilly areas, the land area was larger than in other parts of the city state. This method of measuring land area—based on what could be done with it—is quite different to the Roman system and largely emerged from a method of equitable taxation by which those with poorer land could be taxed fairly alongside those with more fertile land.

This Greek concept of land measurement was later followed by the Anglo-Saxons in England with their use of the "hide" as a measure of land. This measurement was used in the *Domesday Book* in 1086 and continued until the end of the twelfth century. Traditionally, it was thought that a hide consisted of the land needed to support 10 families, because it is used instead of the term *terra x familiarum* (land of 10 families) in the Anglo-Saxon version of Bede's ecclesiastical history. In Scotland during the same period, the term "groatland" was used to describe the land that could be rented for a particular coin—in this case, a groat. It would represent a larger area for poorer agricultural land than for richer land.

Medieval Era

By medieval times, in Europe and especially England, the terms of measuring land were standardized, and these tended to follow the Roman measurements of a "perch," a "rood," and an "acre." In spite of these measures (although the "hide" was being phased out), there were other measures including the "carucate," which covered the land that an eight-ox team could plough in a year (approximately 120 acres); a "virgate," which covered land that could be ploughed by two oxen in a year (about 30 acres); and a "bovate," which covered the land that could be ploughed by a single ox in a year. There was also an area known as a "knight's fee," which was the land expected to be able to produce a single armed soldier in times of war. Although early in medieval England, an acre was supposed to be the land that could be ploughed in single day, by late medieval times, it had been formalized as 4840 square yards.

Other Systems of Measuring Area

Elsewhere in the world, many other places had their own system of measuring area. The Chinese had a system based on the *li* (7.9 square yards), the *fen* (10 *li*), the *mu* (10 *fen*), the *shi* (10 *mu*), and the *qing* (10 *shi*). The Japanese also had a system of measurement by *tsubo*, which covered the land that was the same size as two tatami mats (about 3.306 square meters). In Korea, there is a similar measure called the *pyeong*, which covers 3.3058 square meters. These measures are generally used to measure the size of rooms and buildings rather than large areas of land. The *tsubo* and the *pyeong* are both still used in the twenty-first century to help describe the size of houses or apartments for sale, in the same way as the term "square" is used by Australian estate agents (approximating to 100 square feet, or 9.29 square meters).

Metric System

The metric system was devised during the 1790s following the French Revolution in an attempt to standardize measurements and it was adopted by the French after Napoleon Bonaparte came to power in 1799. It focuses on the meter as the main measurement of length, and the square meter as the measurement of area. This is used throughout most of the world in the twenty-first century.

Further Reading

Anderton, Pamela. *Changing to the Metric System*. London: Her Majesty's Stationery Office, 1965.

Balchin, Paul N., and Jeffrey L. Kieve. *Urban Land Economics*. New York: Palgrave Macmillan, 1985.

Boyd, Thomas D. "Urban and Rural Land Division in Ancient Greece." *Hesperia* 50, no. 4 (October–December 1981).

Kanda, James. "Methods of Land Transfer in Medieval Japan." *Monumenta Nipponica* 33, no. 4 (Winter 1978).

Snooks, Graeme D., and John McDonald. *Domesday Economy: A New Approach to Anglo-Norman History*. Oxford, England: Clarendon Press, 1986.

Walthew, C. V. "Possible Standard Units of Measurement in Roman Military Planning." *Britannia* 12 (1981).

JUSTIN CORFIELD

Units of Length

Category: Space, Time, and Distance.
Fields of Study: Algebra; Geometry; Measurement; Number and Operations; Representations.
Summary: Numerous units of length exist and are used according to the distance measured.

Measuring length or distance has been necessary as far back as the oldest hunter-gatherer peoples in order to perform necessary tasks, such as traveling and finding or hunting food. Many of the first units of length were derived from bodily measurements. Modern units of length can broadly be divided into two categories: the U.S. customary system and the international system. The U.S. customary system is more commonly known as the "American system." The International system is more commonly known as the "metric system." The basic unit of length in the American system is the foot, while the basic unit of length in the metric system is the meter. The American system is used more often in the United States, while the metric system is more common in other parts of the world. Scientific journals almost always report measurements in metric units. The exact values of length measurements depend on the units chosen but certain constants (like π) that are fundamental to related measurements (like circumference) are unitless.

American System

The American system of lengths is similar to the British imperial system from which the American system takes its historical roots. The basic unit of measurement is the foot (ft), which originally was set to be the length of an adult man's foot. Each foot is approximately 0.3048 meters. Smaller distances in the American system are typically measured in inches (in) or less commonly in mils. There are 12 inches in a foot and 1000 mils in an inch. Rather than mils, it is much more common to use fractions of an inch to obtain additional accuracy in the American system. Longer distances in the American system are usually measured in yards (yd) or miles (mi). There are three feet in a yard and 1760 yards in a mile (5280 feet in a mile).

Metric System

The meter was originally established in France as one ten-millionth of the distance from the Earth's equator to the North Pole along the meridian passing through Paris. However, in 1983 it was defined as 1/299,792,458 of the distance traveled by light in a second in a vacuum. Smaller units of length in the metric system are often measured in centimeters (cm), millimeters (mm), micrometers (μm, also known as the *micron*), and nanometers (nm). There are 100 centimeters in a meter and 1000 millimeters in a meter. Similarly, there are 1 million micrometers in a meter and 1 billion nanometers in a meter. Longer distances are usually measured in kilometers (km). There are 1000 meters in a kilometer, which are sometimes referred to as "klicks" in the military. The fermi and the angstrom are also units of length in the metric system, though they are not officially part of the international system. There are 10^{15} fermis in a meter and 10 trillion angstroms in a meter. Because of their small length, the fermi and the angstrom are best suited for very small distances. Less common units of length in the metric system include the decimeter (one-tenth of a meter), picometer (10^{-12} meters), decameter (10 meters), megameter (1 million meters), gigameter (1 billion meters), and petameter (10^{15} meters).

Atomic and Astronomic Measurements

Atomic measurements are also given in terms of either Planck length or the Bohr radius. The Planck length is defined in terms of Planck's constant, the gravitational constant, and the speed of light in a vacuum. The result is that the Planck length is based entirely on universal constants rather than human constructs, such as the second. A Planck length is approximately 1.61625×10^{-35} meters. The Bohr radius is defined as the expected distance between the nucleus of a hydrogen atom and its electron in the Bohr model of the atom. The Bohr radius is approximately 5.29177×10^{-11} meters.

Astronomical distances are typically given in terms of light-years, astronomical units, or parsecs. The light-year is defined as the distance light travels in a vacuum in a Julian year (365.25 days). The light year is approximately 9,460,730,472,581 kilometers or 5,878,630,000,000 miles. Distances such as the light-second, the light-minute, and the light-month are defined analogously to the light-year. The astronomical unit is defined as the average distance between the Earth and the sun, approximately 149,597,871 kilometers or 92,955,807 miles. The parsec is defined in terms of the astronomical unit and an angle with measure one arc second. The imaginary right triangle that defines the parsec has one angle with measure one arc second. The opposite side of the triangle from this angle has length equal to one astronomical unit. The length of the adjacent side to this angle is defined as a parsec and can be derived using basic trigonometry. There are approximately 3.26 light-years in a parsec.

Other Measurements

There are a number of units of length that are based on the American system and still in use in certain professions in the twenty-first century. A furlong is often used in horse racing and is defined as one-eighth of a mile (220 yds). The hand is a unit of length used to describe the height of a horse and is equivalent to four inches. Rods (5.5 yds) and chains (66 ft) are often used in surveying. A fathom is often used to measure the depth of water and is equal to six feet. A nautical mile is approximately equal to one minute of latitude. Thus, there are 1872 meters (approximately 6076 feet) in a nautical mile. Fathoms and nautical miles are often used by mariners.

There are also a number of archaic units of length that may be familiar to the reader, most significantly the cubit (1.5 ft) and the league. The dimensions of Noah's Ark as well as other Biblical artifacts are given in cubits. The league has several different values, however the most common is the distance that a person can walk in an hour (approximately three miles). The league was featured in the title of Jules Verne's *Twenty Thousand Leagues Under the Sea*.

Further Reading

Glover, Thomas G. *Measure for Measure*. Littleton, CO: Sequoia Publishers, 1996.

Hopkins, Robert A., *The International (SI) System and How It Works*. Tarzana, AK: AMJ Publishing, 1975.

Liflander, Pamela. *Measurements & Conversions*. Philadelphia: Running Press, 2003.

Wildi, Theodore. *Metric Units and Conversion Charts*. Hoboken, NJ: Wiley-IEEE Press, 1995.

Young, Hugh D., et al. *University Physics With Modern Physics*. 12th ed. Boston: Addison-Wesley, 2007.

ROBERT A. BEELER

Units of Mass

Category: Space, Time, and Distance.
Fields of Study: Algebra; Geometry; Measurement; Number and Operations; Representations.
Summary: A variety of measurement systems have been used throughout history to measure weight and mass.

Throughout history, there have been many ways of measuring mass. Until modern times, these methods were those used to measure what was known as "weight." A number of ways of assessing weight existed in prehistoric times. The Sumerians used a system similar to that later used throughout the ancient Middle East, with 180 grains making a shekel (or *gin*), and 60 of these forming a pound (or *ma-na*), and 600 of these making a load (or *gun*). A wall painting from ancient Egypt, dating from 1285 B.C.E., shows the god Anubis weighing the heart of Hunefer using scales, indicating that the Egyptians had a system of using weights and measures. There were, however, slight differences between the Middle Kingdom and the New Kingdom in Egypt.

Greeks and Romans

The Greeks, with the extensive use of coinage, used a scale that was based on the barley corn but it was actually more fixed on the weight of individual coins. The Romans adapted the Greek system for their own use, with the basic measure of an *uncia* (or ounce). Twelve of these made up one *as*, with different names were given to parts of an *as*: *quadrans* were a quarter of an *as* and *semis* were half an *as*.

Middle Ages

During the Middle Ages in Europe, there were a number of measures that were used for a variety of purposes. For apothecaries, jewelers, and the making of coins, there were "grains," "scruples," and "drams." Two systems were heavily used in Western Europe. The Troy weights, named after the French city of Troyes, were based on the troy ounce (the name "ounce" coming from the Roman "uncia"). By contrast in England, until 1526, there was the Tower ounce, which was slightly lighter than its continental measure (18.75 dwt/pennyweight, rather than the Troy ounce which was 20 dwt). For both measures, 12 ounces made up a pound. In

England, eight pounds equaled a "butcher's stone," and 12 pounds a "mercantile stone." The larger measurements were in tons, which consisted of 2240 pounds—now known as a "long ton." The United States later adopted a measure in which 2000 pounds equals a "short ton."

Standardization

An attempt to standardize the measurement of mass started in France, which, on December 10, 1799, passed a new law establishing a kilogram that consisted of 18,827.15 grains, although the kilogram had already beem used for the previous four years. It was defined as being a 1 cubic decimeter of distilled water at 4 degrees centigrade, its maximum density. This standardization led to the metric system, and in turn it led to the introduction of what became known as the International System of Units (SI units).

The SI units define mass in kilograms as the base unit. It is almost exactly the same as one liter of water, although the exact measure is the same as a piece of platinum-iridium alloy, which is called the International Prototype Kilogram and is stored in a vault in France. An anomaly meant that it was the only base unit with an SI prefix "kilo-" (meaning thousand). There are a number of multiples and submultiples, but only some are commonly used. A one-thousandth part of a gram is called a "milligram," a millionth part called a "microgram," and 10^{-9} g called a "nanogram." Going the other way, although there are terms such as a "zettagram" (10^{21} g) and a "yottagram" (10^{24} g), these are rarely used. One curiosity is that instead of using the term "megagram," for 1000 kg, the term "tonne" is used; its spelling denoting its difference from the pre-decimal "ton." It has been relatively easy to convert from the old "imperial" system of pounds to the metric SI system, with one kilogram essentially being 2.2 pounds. Most of the world uses kilograms in the twenty-first century, the United States being a prominent exception.

Throughout Europe, there were regional varieties and customary names. Scotland was divided between using the "Troy" measures, and the "Tron" measures, the latter being used in Edinburgh—the system was standardized in 1661. The Portuguese used a system maintained at a national level and was based on the *onca* (ounce), with 16 of these making an *arratel* (pound), 128 *arrateis* making a *quintal*, and 1728 making a *tonelada*. These Portuguese measures, also used in Brazil, were abandoned when both countries adopted the metric system: Portugal and its colonies (or overseas provinces) in 1852, and Brazil 10 years later. The Russians also had their own system, which had emerged from that used by the Mongols—although Peter the Great (r. 1682–1725) overhauled the system and used one based on the English system.

Asia

Elsewhere in the world, there were many other systems of measuring mass. The Chinese used a system with 1000 *cash* making a *tael*, and ten *taels* equaling a *catty*, and 100 of those making up a *picul*. The Japanese system relied on the *momme* (about 3.75 g), with 100 of these forming a *hyakume*, 160 of them making one *kin*, and 1000 of them equaling one *kan*. The *momme* is still used as a measure of mass in the pearling industry, which is still dominated by Japan.

Further Reading

Anderton, Pamela. *Changing to the Metric System*. London: Her Majesty's Stationery Office, 1965.

Fenna, Donald. *A Dictionary of Weights, Measures, and Units*. New York: Oxford University Press, 2002.

Grayson, Michael A. *Measuring Mass: From Positive Rays to Proteins*. Darby, PA: Chemical Heritage Foundation, 2005.

Moody, Ernest A. *The Medieval Science of Weights*. Madison: University of Wisconsin Press, 1960.

Zupko, Ronald Edward. "Medieval Apothecary Weights and Measures: The Principal Units of England and France." *Pharmacy in History* 32 (1990).

Justin Corfield

Units of Volume

Category: Space, Time, and Distance.
Fields of Study: Algebra; Geometry; Measurement; Number and Operations; Representations.
Summary: A variety of units are used to measure volume throughout the world.

Measuring of volume intrigued many scientists in the ancient world. For the most part, crops, stones, and other items were measured by weight rather than volume because of the relative ease of doing so—especially given the irregular shapes of many items. For solid items with irregular shapes, it seemed far too complicated to work out their volume, even if this could be done with any degree of accuracy. This notion changed dramatically with the ideas that have been attributed to the famous Greek mathematician and inventor Archimedes of Syracuse (c. 287–212 B.C.E.). The tale of Archimedes is that he was given the task of determining the purity of the gold used to create the crown of King Hiero II—the king was worried that silver or base metals might have been used in its manufacture and were cleverly disguised. Pondering the problem while getting into a bathtub, Archimedes, according to the story, noticed that the water rose and the amount it rose was equal to the size of the parts of his body that were submerged. This led Archimedes to deduce that water could be used to measure the volume of a particular item, such as the king's crown. It could then be weighed against a block of pure gold of the same volume. It is said that when he realized that this could be done, Archimedes shouted "Eureka!" ("I have found it!") and ran through the streets to tell everybody of his discovery, forgetting that he had not put his clothes on.

Whether or not the story of Archimedes is actually true—and some historians doubt its veracity, although Galileo stated that he believed that it might well be true—the story does illustrate the use of fluid displacement, which can be used to easily measure the volume of irregularly shaped objects. This method does not seem to have been known before the Greeks. Certainly, the ancient Egyptians had major problems working out volume and there are complicated equations and formulae on the Rhind Mathematical Papyrus, which dates to about 1700 B.C.E., illustrating that the Egyptians were already grappling with the subject.

The Romans had two systems for recording the measurement of volume. The first and most often used was for liquid measures and was based on a sextarius (from *sester*), which is roughly 0.54 liters. Six *sesters* made up one *congius*; four of these made up one *urn*; and two *urns* make one *amphora*. For dry measures, although a *sester* was still used and was of equivalent size, eight *sesters* equaled a gallon; two gallons made up one modius (also called "peck"); and three of these one quadrantal (also called "bushel").

English
In use from the Middle Ages, the English ended up with an extremely complicated system of measuring volume, which was formalized as Imperial Measurements. The smallest measure was a "mouthful," with two of those making a "pony," two ponies making a "jack," two jacks making one "gill," two gills making one "cup," and two of those making a "pint." The system continued with two pints making a "quart" (from "quarter gallon"), with two quarts equaling one "pottle," and two pottles making a "gallon." The next levels of measurements were "pecks," "kennings," "bushels," "strikes," "coombs," "hogsheads," and "butts" (also called "pipes"). A slightly different scale was used to measure wine and beer. Even when British adopted the metric system in 1965, some of the old terminology (and measures) were still used, especially pints (for milk) and gallons (for gasoline). A bushel is also the standard measurement for wheat and some other items in agriculture.

Metric System
Although the English had numerous terms, medieval and early modern France had a vast range of measures of volume, which varied from one part of France to another, most arising for customary reasons. After the French Revolution, the new government sought to standardize all systems of measurement, including volume. This process saw the introduction, under Napoleon Bonaparte, of metrication, and in turn it led to the International System of Units (SI units).

The original metric system had liters (or litres) as the measure of volume, and from 1901 until 1964, it was defined as being the volume of one kilogram of pure water heated to 4 degrees Centigrade and measured under a pressure of 760 millimeters of mercury.

When it came to devising the SI units, the liter was dropped as a measure, and the official measurement was in cubic meters. The difference is not significant in all but scientific terms, although liters continued to be used by many people throughout the world.

Gas Volume

While it was possible to measure the volumes of liquids easily (and also of solid objects by measuring the displacement of a water of a similar quantity), the measuring of the volume of gas has long posed a problem. The problem was solved by the British civil engineer Samuel Clegg (1781–1861), who had worked on natural gas flues and was able to design a dry meter and then a water meter, which were able to measure the amount of gas used by consumers. This invertion helped the gas industry in Britain—and later in other countries—measure gas and thereby charge customers based on usage.

Further Reading

Falkus, M. E. "The British Gas Industry Before 1850." *The Economic History Review New Series* 20, no. 3 (December 1967).

Hirshfeld, Alan. *Eureka Man: The Life and Legacy of Archimedes.* New York: Walker, 2009.

Jaeger, Mary. *Archimedes and the Roman Imagination.* Ann Arbor: University of Michigan Press, 2008.

Lawn, Richard E., and Elizabeth Prichard. *Measurement of Volume.* Cambridge, England: Royal Society of Chemistry, 2003.

Justin Corfield

Vectors

Category: History and Development of Curricular Concepts.
Fields of Study: Communication; Connections; Measurement; Number and Operations.
Summary: Vectors express magnitude and direction, and have applications in physics and many other areas.

There are some quantities, like time and work, that have only a magnitude (also called "scalars"). If one says the time is 6 a.m., it is adequate. When discussing velocity or force, however, then magnitude is not enough. If a particle has a velocity of five meters per second, this is not sufficient information because the direction of movement is unknown. Quantities that require both a magnitude and a sense of direction for their complete specifying are called "vectors." Pilots use vectors to compensate for wind to navigate airplanes, sport analysts use vectors to model dynamics, and physicists use vectors to model the world.

History and Development of Vectors

The term "vector" originates from *vectus*, a Latin word meaning "to carry." However, astronomy and physical applications motivated the concept of a vector as a magnitude and direction. Aristotle recognized force as a vector. Some historians question whether the parallel law for the vector addition of forces was also known to Aristotle, although they agree that Galileo Galilei stated it explicitly and it appears in the 1687 work *Principia Mathematica* by Isaac Newton. Aside from the physical applications, vectors were useful in planar and spherical trigonometry and geometry. Vector properties and sums continue to be taught in high schools in the twenty-first century.

The rigorous development of vectors into the field of vector calculus in the nineteenth century resulted in a debate over methods and approaches. The algebra of vectors was created by Hermann Grassmann and William Hamilton. Grassmann expanded the concept of a vector to an arbitrary number of dimensions in his book *The Calculus of Extension*, while Hamilton applied vector methods to problems in mechanics and geometry using the concept of a "quaternion." Hamilton spent the rest of his life advocating for quaternions. James Maxwell published his *Treatise on Electricity and Magnetism* in which he emphasized the importance of quaternions as mathematical methods of thinking, while at the same time critiquing them and discouraging scientists from using them. Extending Grassman's ideas, Josiah Gibbs laid the foundations of vector analysis and created a system that was more easily applied to physics than Hamilton's quaternions. Oliver Heaviside independently created a vector analysis and advocated for vector methods and vector calculus. Mathematicians such as Peter Tait, who preferred quaternions, rejected the methods of Gibbs and Heaviside. However, their methods were eventually accepted and they are taught as part of the field of linear algebra. The quaternionic method of Hamilton remains extremely useful in the twenty-first century. Vector calculus is fundamental in understanding fluid dynamics, solid mechanics, electromagnetism, and in many other applications.

During the nineteenth century, mathematicians and physicists also developed the three fundamental theorems of vector calculus, often referred to in the twenty-first century as the "divergence theorem," "Green's theorem," and "Stokes' theorem." Mathematicians with diverse motivations all contributed to the development of the divergence theorem. Michael Ostrogradsky studied the theory of heat, Simeon Poisson studied elastic bodies, Frederic Sarrus studied floating bodies, George Green studied electricity and magnetism, and Carl Friedrich Gauss studied magnetic attraction. The theorem is sometimes referred to as "Gauss's theorem." George Green, Augustin Cauchy, and Bernhard Riemann all contributed to Green's theorem, and Peter Tait and James Maxwell created vector versions of Stokes theorem, which was originally explored by George Stokes, Lord Kelvin, and Hermann Hankel. Undergraduate college students often explore these theorems in a multivariable calculus class.

The concept of a space consisting of a collection of vectors, called a "vector space," became important in the twentieth century. The notion was axiomatized earlier by Jean-Gaston Darboux and defined by Giuseppe Peano, but their work was not appreciated at the time. However, the concept was rediscovered and became important in functional analysis because of the work by Stefan Banach, Hans Hahn, and Norbert Wiener, as well as in ring theory because of the work of Emmy Noether. Vector spaces and their algebraic properties are regularly taught as a part of undergraduate linear algebra.

Mathematics

A vector is defined as a quantity with magnitude and direction. It is represented as a directed line segment with the length proportional to the magnitude and the direction being that of the vector. If represented as an array, it is often represented as a row or column matrix. Vectors are usually represented as boldface capital letters, like A or with an arrow overhead: $\vec{A}$.

The Triangle Law states that while adding, "if two vectors can be represented as the two sides of a triangle taken in order then the resultant is represented as the closing side of the triangle taken in the opposite order" (see Figure 1).

Any vector can be split up into components, meaning to divide it into parts having directions along the coordinate axes. When added, these components return the original vector. This process is called "resolution into components" (see Figure 2). Clearly, this resolution cannot be unique as it depends on the choice of coordinate axes. However, for a given vector and specified coordinate axes, the resolution is unique. When two vectors are added or subtracted, these components along a specific axis simply "add up" (like 2 + 2 = 4 or 7 − 2 = 5) but the original vectors do not, which follow the rule of vector addition that can be obtained by the Parallelogram Law of Vector Addition. Vector addition is commutative and associative in nature.

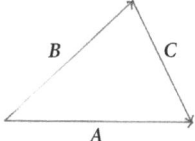

Figure 1. **B** and **C** add up to **A**.

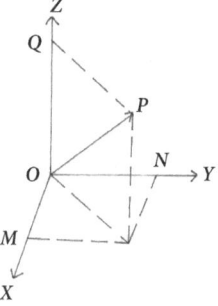

Figure 2. **OP** can be split into mutually perpendicular components **OM**, **ON**, and **OQ**.

Multiplication for vectors can be of a few types:

1. For scalar multiplication (multiplication by a quantity that is not a vector), each component is multiplied by that scalar. Vector multiplication by a scalar is commutative, associative, and distributive in nature.
2. For the multiplication of two vectors, one can obtain both a scalar (dot product) or a vector (cross product). For a cross product the resultant lies in a plane perpendicular to the plane containing the two original vectors. Dot product is both commutative and distributive. But cross product is neither commutative nor associative in nature because the result is a vector and depends on the direction.

Applications

Theoretical sciences have a wide spread of applications of vectors in nearly all fields:

- *Obtaining components*: Occasionally, one needs a part (or component) of a vector for a given purpose. For example, suppose a rower intends to cross over to a point on the other side of a river that has a great current. The rower would be interested to know if any part of that current could help in any way to move in the desired direction. To find the component of the current's vector along any specified direction, take the dot product of that vector with a unit vector (vector of unit magnitude) along the specified direction. This method is of particular importance in studying of particle dynamics and force equilibria.
- *Evaluating volume, surface, and line integrals*: In many problems of physics, it is often necessary to shift from either closed surface integral (over a closed surface that surrounds a volume) to volume integral (over the whole enclosed volume), or from closed line integral (over a loop) to surface integrals (over a surface). To accomplish these shifts, it is often very useful to apply two fundamental theorems of vector calculus, namely Gauss's divergence theorem and Stokes's theorem, respectively.
- *Particle mechanics*: In the study of particle mechanics, vectors are used extensively. Velocity, acceleration, force, momentum, and torque all being vectors, a proper study of mechanics invariably involves extensive applications of vectors.
- *Vector fields*: A field is a region over which the effect or influence of a force or system is felt. In physics, it is very common to study electric and magnetic fields, which apply vectors and vectorial techniques in their description.

Further Reading

Katz, Victor. "The History of Stokes' Theorem." *Mathematics Magazine* 52, no. 3 (1979).

Matthews, Paul. *Vector Calculus*. Berlin: Springer, 1998.

Stroud, K. A., and Dexter Booth. *Vector Analysis*. New York: Industrial Press, 2005.

Abhijit Sen

Zero

Category: History and Development of Curricular Concepts.
Fields of Study: Communication; Connections; Number and Operations.
Summary: The concept of zero took time to be accepted and was explicitly rejected when first introduced to Greek and Roman culture.

Numbers initially served to count property, such as livestock. The numbers needed to count 1, 2, 3, 4, ... became known as "counting" or "natural" numbers. The number zero is not found among these because one cannot count zero objects. Early civilizations existing over millennia used numbers only to count and so had no need for zero. The word "zero" has various linguistic origins: the French *zéro* and Venetian *zero*, which likely evolved from the Italian *zefiro*. This word came in turn from Arabic *sifr*, meaning "zero or nothing," derived from word *safira*, meaning "it was empty."

Early Development

The ancient Babylonians first introduced zero. With a base-60 system and initially two symbols (a wedge to represent "1" and a double wedge to represent 10), the Babylonians left empty spaces between groups of symbols. The fact that the spaces were not standardized in length made it difficult at times to distinguish between numbers because place value could not always be determined. To remedy this situation, the Babylonians developed zero but the zero was not a number in and of itself. It was rather a placeholder used to denote place values that had been skipped.

Independently and across the ocean, the Mayans developed a base-20 number system that included zero. Here, zero was used as a number to mean the absence of something. Zero also appeared in the Mayans' calendar. There was a year zero, and each month had a day zero in it as well. Because of the vast distance between the Mayans and the old world, Mayans' use and understanding of zero did not spread to these other areas.

Rejection by the Greeks and Romans

Despite the Babylonians use of zero, the Greeks and Romans initially rejected its use. Zero was considered dangerous spiritually as it represented the opposite of god and unity. It was associated with the void and chaos. Mathematically, zero presented many dilemmas. While any of the natural numbers (1, 2, 3, 4, . . .) when added to itself yields a larger number, zero added to itself does not. This characteristic violated Archimedes's principal that repeatedly adding a number to itself tends to a sum that is infinitely large. Additionally, a natural number plus any other natural number yields a sum larger than the initial natural number but again zero added to a natural number does not yield a number larger than the original natural number. Finally, multiplication of any number by zero yields zero and division by zero was outside the acceptable norms for these civilizations. The Greeks, known for geometry, often associated geometric figures to the natural numbers but zero could be associated with no figure. They preferred to reject zero as a number altogether.

Zero in India

Indian mathematicians in the fifth century C.E. took ideas from the Babylonians, including the concept of zero. They treated zero as a number that was found in the number line between −1 and 1. They also introduced negative numbers and, in 700, Brahmagupta introduced the idea that $1/0 = \infty$. Thus, infinity and unity depend upon the void and chaos. This idea was troubling to many civilizations, and the Hindu-Arabic numerals commonly used through the twenty-first century were not fully accepted until Leonardo de Pisa (also known as Fibonacci) introduced them to the Western world in his 1202 work *Liber Abaci*. One of the earliest recorded references to the mathematical impossibility of assigning a value to 1/0 occurred in George Berkeley's 1734 work *The Analyst*, which criticizes the foundations of calculus.

Calendars

Zero also caused confusion with the calendar system. Dionysius' calendar, created in 525 C.E., introduced the notation of BC and AD. However, it did not include a year zero. Thus, 1 BC is followed by 1 AD. This omission of zero causes confusion into the twenty-first century. Consider a person born in 1 AD. This person would have to go through 1, 2, 3, 4, 5, 6, 7, 8, 9, and 10 to have lived 10 years, and a new decade would begin at the end of this first decade (10 years). That is, it would begin in 11. Thus, the next decade would begin in 21. The first century would end in 100, and the new one would begin in 101. Thus, the twenty-first century technically began in 2001, not in 2000 when most everyone celebrated it. This confusion rears its head at the start of every decade and century all a result of the omission of a year zero.

Division by Zero

One way in which mathematicians interpret division by zero is to reframe division in terms of other arithmetic operations. Using standard rules for arithmetic, division by zero is undefined, since division is defined to be the inverse operation of multiplication. While division by zero cannot reasonably be resolved with real numbers and integers, it can be defined using other algebraic structures or analytical extensions.

Zero in the Physical Sciences

Zero is an important value for many physical quantities or measurements. In some cases, zero means "nothing" or an absence of the characteristic, such as in most units of length and mass. However, in some cases, zero represents an arbitrarily chosen starting point for counting or measuring, such as in the Fahrenheit and Celsius temperature scales (though on the Kelvin scale, zero is the coldest possible temperature that matter can reach).

Other more advanced examples can be found in chemistry and physics. Zero-point energy is the lowest possible energy that a quantum mechanical physical system may possess. This energy level is called the "ground state" of the system and is important for investigating concepts such as entropy and perfect crystal lattices. Professor Andreas von Antropoff introduced the term "neutronium" for theoretical matter made solely of neutrons. As early as 1926, he redefined the periodic table with the atomic number zero, rather than the standard hydrogen (Atomic Number 1) in the initial position. More recent investigations suggest that the hypothesized element tetraneutron, a stable cluster of four neutrons with no protons or electrons, could have this atomic number zero.

Zero and Computers

In 1997, the naval vessel USS *Yorktown*'s propulsion system was brought to a dead stop by a computer network failure resulting from an attempt to divide by zero. Mathematical operations like these are problematic for computers, leading to various methods to avoid errors. The floating-point standard used in most modern computer processors has two distinct zeroes: a +1 (positive zero) and a −0 (negative zero). They are considered equal in numerical comparisons but some mathematical operations will have different results depending on which zero is used. For example, 1/−0 yields negative infinity, while 1/+1 gives positive infinity, though a "divide by zero" warning is usually issued in either case. Integer division by zero is usually handled differently from floating point, as there is no integer representation for the answer. Some processors generate an exception for integer division by zero, although others will simply generate an incorrect result for the division.

Further Reading

Ifrah, Georges, and Lowell Bair. *From One to Zero: A Universal History of Numbers*. New York: Penguin, 1987.

Kaplan, Robert. *The Nothing That Is: A Natural History of Zero*. Oxford, England: Oxford University Press, 2000.

Seife, Charles. *Zero: The Biography of a Dangerous Idea*. New York: Penguin Books, 2000.

Lidia Gonzalez

Resource Guide

Books

Aaboe, Asger. *Episodes From the Early History of Mathematics*. Washington, DC: Mathematical Association of America, 1975.

Adrian, Yeo. *The Pleasures of Pi and Other Interesting Numbers*. Singapore: World Scientific Publishing, 2006.

Agresti, A. *Categorical Data Analysis*. Hoboken, NJ: Wiley, 2002.

Aho, A. V., J. E. Hopcrotf, and J. D. Ullman. *The Design and Analysis of Computer Algorithms*. Reading, MA: Addison-Wesley, 1976.

Albert, Jim, and Jay Bennett. *Curve Ball: Baseball, Statistics, and the Role of Chance in the Game*. New York: Springer-Verlag, 2001.

Ascher, Marcia. *Mathematics Is Everywhere: An Exploration of Ideas Across Cultures*. Princeton, NJ: Princeton University Press, 2002.

Ball, W. W. Rouse. *A Short Account of the History of Mathematics*. New York: Sterling Publishing Company, 2001.

Barnett, Raymond, Michael Ziegler, and Karl Byleen. *Calculus for Business, Economics, Life Science, and Social Science*. Upper Saddle River, NJ: Prentice-Hall, 2005.

Baumohl, Bernard. *The Secrets of Economic Indicators: Hidden Clues to Future Economic Trends and Investment Opportunities*. 2nd ed. Upper Saddle River, NJ: Pearson Education, 2008.

Beckmann, Petr. *A History of π (Pi)*. New York: Barnes & Noble, 1971.

Behrends, Ehrhard. *Five-Minute Mathematics*. Providence, RI: American Mathematical Society, 2008.

Bell, Eric Temple. *Men of Mathematics*. New York: Simon & Schuster, 1937.

Bennett, Jay, and James Cochran. *Anthology of Statistics in Sports*. Philadelphia, PA: Society for Industrial and Applied Mathematics, 2005.

Berggren, Lennart, Jon Borwein, and Peter Borwein. *Pi: A Source Book*. New York: Springer-Verlag, 1997.

Berlekamp, Elwyn R., John H. Conway, and Richard K. Guy. *Winning Ways for Your Mathematical Plays*. Natick, MA: AK Peters, 2001.

Blackwell, William. *Geometry in Architecture*. Hoboken, NJ: Wiley, 1984.

Blatner, David. *The Joy of π*. New York: Walker & Co., 1997.

Blue, Ron, and Jeremy White. *The New Master Your Money: A Step-by-Step Plan for Gaining and Enjoying Financial Freedom*. Chicago: Moody, 2004.

Blum, Raymond. *Mathemagic*. New York: Sterling Publishing, 1992.

Bodie, Zvi, Alex Kane, and Alan Marcus. *Investments*. Chicago, IL: McGraw-Hill/Irwin, 2008.

Borwein, Jonathan, and Peter Borwein. *A Dictionary of Real Numbers*. Pacific Grove, CA: Brooks/Cole Publishing Co., 1990.

Boyer, C. B. *A History of Mathematics*. Hoboken, NJ: Wiley, 1968.

Boyer, C. B. *The History of the Calculus and Its Conceptual Development*. New York: Dover Publications, 1949.

Brealey, Richard A., Stewart C. Myers, and Franklin Allen. *Principles of Corporate Finance*. 9th ed. New York: McGraw-Hill, 2008.

Bressoud, David. *The Queen of the Sciences: A History of Mathematics*. Chantilly, VA: The Teaching Company, 2008.

Broverman, Samuel A. *Mathematics of Investment and Credit*. Winsted, CT: ACTEX Publications, 2008.

Burkett, Larry, and Brenda Armstrong. *Making Ends Meet: Budgeting Made Easy*. Gainesville, GA: Crown Financial Ministries, 2004.

Burton, David M. *The History of Mathematics: An Introduction*. New York: McGraw-Hill, 2005.

Calinger, Ronald. *A Contextual History of Mathematics*. Upper Saddle River, NJ: Prentice-Hall, 1999.

Clagett, Marshall. *Archimedes in the Middle Ages*. Madison: University of Wisconsin Press, 1964.

Closs, Michael. *A Survey of Mathematics Development in the New World*. Ottawa: University of Ottawa, 1977.

Closs, Michael, ed. *Native American Mathematics*. Austin: University of Texas Press, 1986.

Coe, Michael D. *Breaking the Maya Code*. New York: Thames and Hudson, 1992.

Copeland, Thomas E., J. Fred Weston, and Kuldeep Shastri. *Financial Theory and Corporate Policy*. 4th ed. Upper Saddle River, NJ: Pearson Education, 2005.

Cullen, Christopher. *Astronomy and Mathematics in Ancient China: The Zhou Bi Suan Jing*. Cambridge, England: Cambridge University Press, 1996.

Cuomo, Serafina. *Ancient Mathematics*. London: Routledge, 2001.

Davenport, Harold. *The Higher Arithmetic: An Introduction to the Theory of Numbers*. Cambridge, England: Cambridge University Press, 1999.

Davis, Morton D. *The Math of Money: Making Mathematical Sense of Your Personal Finances*. New York: Copernicus, 2001.

De Mestre, Neville. *The Mathematics of Projectiles in Sport*. Cambridge, England: Cambridge University Press, 1990.

Devlin, Keith. *The Math Gene: How Mathematical Thinking Evolved and Why Numbers Are Like Gossip*. New York: Basic Books, 2001.

———. *The Unfinished Game: Pascal, Fermat, and the Seventeenth-Century Letter That Made the World Modern*. New York: Basic Books, 2008.

Drobot, Stefan. *Real Numbers*. Upper Saddle River, NJ: Prentice-Hall, 1964.

Dudley, Underwood. *Numerology or What Pythagoras Wrought*. Washington, DC: Mathematical Association of America, 1997.

Eastway, Rob, and John Haigh. *Beating the Odds: The Hidden Mathematics of Sport*. London: Robson Books, 2007.

Eglash, Ron. *African Fractals: Modern Computing and Indigenous Design*. New Brunswick, NJ: Rutgers University Press, 1999.

Eves, Howard. *An Introduction to the History of Mathematics*. New York: Saunders College Publishing, 1990.

Flegg, G. *Numbers: Their History and Meaning*. New York: Schocken Books, 1983.

Friberg, Jöran. *Unexpected Links Between Egyptian and Babylonian Mathematics*. Singapore: World Scientific Publishing Co., 2005.

Friedman, Arthur. *World of Sports Statistics: How the Fans and Professionals Record, Compile, and Use Information*. New York: Athenaeum, 1978.

Fries, Christian. *Mathematical Finance: Theory, Modeling, Implementation*. Hoboken, NJ: Wiley, 2007.

Frumkin, Norman. *Guide to Economic Indicators*. Armonk, NY: M. E Sharpe, 2000.

Gamow, George. *One, Two, Three... Infinity*. New York: Viking Press, 1947.

Gardner, David, and Tom Gardner. *The Motley Fool Personal Finance Workbook: A Foolproof Guide to Organizing Your Cash and Building Wealth*. New York: Fireside Books, 2003.

Gardner, Martin. *Mathematics, Magic and Mystery*. New York: Dover, 1956.

Gay, Timothy. *The Physics of Football*. New York: HarperCollins, 2005.

Gerdes, Paulus. *Geometry From Africa: Mathematical and Educational Explorations*. Washington, DC: Mathematical Association of America, 1999.

Gillings, R. J. *Mathematics in the Time of the Pharaohs*. New York: Dover Publications, 1982.

Gutstein, Eric, and Bob Peterson, eds. *Rethinking Mathematics: Teaching Social Justice by the Numbers*. Milwaukee, WI: Rethinking Schools, 2005.

Hadamard, Jacques. *A Mathematician's Mind*. Princeton, NJ: Princeton University Press, 1996.

Hardy, G. H. *A Mathematician's Apology*. Cambridge, England: Cambridge University Press, 1941.

Henry, Granville C. *Logos: Mathematics and Christian Theology*. Lewisburg, PA: Bucknell University Press, 1976.

Hersh, Rueben. *What Is Mathematics, Really?* New York: Oxford University Press, 1997.

Hoyle, Joe Ben, Thomas F. Schaefer, and Timothy S. Doupnik. *Fundamentals of Advanced Accounting.* New York: McGraw-Hill, 2010.

Kalbfleisch, John D., and Ross L. Prentice. *The Statistical Analysis of Failure Time Data.* Hoboken, NJ: Wiley, 2002.

Katz, Victor J., ed. *Mathematics of Egypt, Mesopotamia, China, India, and Islam: A Sourcebook.* Princeton, NJ: Princeton University Press, 2007.

Kellison, Stephen G. *Theory of Interest.* New York: McGraw-Hill, 2009.

Kimmel, Paul D., Jerry J. Weygandt, and Donald E. Keiso. *Financial Accounting: Tools for Business Decision Making.* Hoboken, NJ: Wiley, 2009.

King, Jerry. *The Art of Mathematics.* New York: Plenum Press, 1992.

Klein, John P., and Melvin L. Moeschberger. *Survival Analysis: Techniques for Censored and Truncated Data.* New York: Springer-Verlag, 1997.

Kline, M., *Mathematical Thought From Ancient to Modern Times.* New York: Oxford University Press, 1972.

Koetsier, T., and L. Bergmans, eds. *Mathematics and the Divine: A Historical Study.* Amsterdam: Elsevier, 2005.

Longe, Bob. *The Magical Math Book.* New York: Sterling Publishing, 1997.

Martzloff, Jean-Claude. *A History of Chinese Mathematics.* New York: Springer-Verlag, 1987.

Moses, Robert P., and Charles E. Cobb, Jr. *Radical Equations: Civil Rights From Mississippi to the Algebra Project.* Boston: Beacon Press, 2001.

Mullis, Darrell, and Judith Handler Orloff. *The Accounting Game: Basic Accounting Fresh From the Lemonade Stand.* Naperville, IL: Sourcebooks, 2008.

Nahin, Paul J. *Dr. Euler's Fabulous Formula.* Princeton, NJ: Princeton University Press, 2006.

Nasar, Sylvia. *A Beautiful Mind: The Life of Mathematical Genius and Nobel Laureate John Nash.* New York: Simon & Schuster, 2001.

Oliver, Dean. *Basketball on Paper: Rules and Tools for Performance Analysis.* Washington, DC: Brassey's, 2004.

Pullan, J. M. *The History of the Abacus.* New York: F. A. Praeger, 1969.

Rafiquzzaman, M. *Fundamentals of Digital Logic and Microcomputer Design.* Hoboken, NJ: Wiley, 2005.

Rudin, W. *Principles of Mathematical Analysis.* New York: McGraw-Hill, 1953.

Salem, Lionel, Frédéric Testard, and Coralie Salem. *The Most Beautiful Mathematical Formulas.* Hoboken, NJ: Wiley, 1992.

Schwarz, Alan. *The Numbers Game: Baseball's Lifelong Fascination with Statistics.* New York: St. Martin's Press, 2004.

Smith, D. E. *History of Mathematics.* Vol. 2. New York: Dover Publications, 1958.

Solow, Daniel. *How to Read and Do Proofs: An Introduction to Mathematical Thought Process.* Hoboken, NJ: Wiley, 1982.

Steen, Lynn A. *On the Shoulders of Giants: New Approaches to Numeracy.* Washington, DC: National Academy Press, 1990.

Sterrett, Andrew. *101 Careers in Mathematics.* Washington, DC: The Mathematical Association of America, 1996.

Suzuki, Jeff. *A History of Mathematics.* Upper Saddle River, NJ: Prentice Hall, 2002.

Taylor, Alan D. *Mathematics and Politics: Strategy, Voting Power, and Proof.* New York: Springer-Verlag, 1995.

van der Waerden, B. L. *Geometry and Algebra in Ancient Civilizations.* Berlin: Springer, 1983.

Venema, G.A. *The Foundations of Geometry.* Upper Saddle River, NJ: Pearson Prentice Hall, 2006.

Weygandt, Jerry J., Paul D. Kimmel, and Donald E. Keiso. *Managerial Accounting: Tools for Business Decision Making.* Hoboken, NJ: Wiley, 2008.

Winkler, Peter. *Mathematical Puzzles: A Connoisseur's Collection.* Natick, MA: AK Peters, 2004.

Wright, Tommy, and Joyce Farmer. *A Bibliography of Selected Statistical Methods and Development Related to Census 2000.* Washington, DC: U.S. Bureau of the Census, 2000.

Yeldham, F. A. *The Teaching of Arithmetic Through Four Hundred Years (1535–1935).* London: G. G. Harrap & Company, 1935.

Yong, L. L., and A. T. Se. *Fleeting Footsteps.* Singapore: Word Scientific Publications, 2004.

Zaslavsky, Claudia. *Africa Counts: Number and Pattern in African Culture.* Chicago: Lawrence Hill Books, 1999.

Zill, D. G. *Calculus with Analytic Geometry.* Boston: Prindle, Weber & Schmidt, 1985.

Journals and Magazines

The AMATYC Review

The American Mathematical Monthly

Association for Women in Mathematics Newsletter

Biometrics
Chance
The College Mathematics Journal
Experimental Mathematics
The Fibonacci Quarterly
Historia Mathematica
IMU-Net
Involve
Journal of Humanistic Mathematics
Journal of Integer Sequences
Journal of Recreational Mathematics
Journal of Statistics Education
Loci
MAA FOCUS
Math Horizons
Mathematics Magazine
Mathematics Teacher
NAM Newsletter
Notices of the American Mathematics Society
The Pentagon
Pi Mu Epsilon Journal
Plus Magazine
PRIMUS
Rose-Hulman Undergraduate Mathematics Journal
SIAM Review
Scholastic Math
Significance
Teaching Children Mathematics
Undergraduate Mathematics and Its Applications

Internet

American Institute of Mathematics
    www.aimath.org
The Algebra Project
    www.algebra.org
AMATYC
    www.amatyc.org
American Mathematical Society
    www.ams.org
American Statistical Association
    www.amstat.org
Association for Women in Mathematics
    www.awm-math.org
CryptoKids
    www.nsa.gov/kids
Datamath Calculator Museum
    www.datamath.org
Illuminations
    illuminations.nctm.org
MacTutor History of Mathematics
    www-history.mcs.st-and.ac.uk
Mathematical Fiction
    http://kasmana.people.cofc.edu/MATHFICT
Math for America
    www.mathforamerica.org
Math Forum
    www.mathforum.com
Math Fun Facts!
    www.math.hmc.edu/funfacts
MathDL
    mathdl.maa.org/mathDL
Mathematical Association of America
    www.maa.org
Mathematical Science Research Institute
    www.msri.org
The Museum of Mathematics
    www.momath.org
National Association of Mathematicians
    www.nam-math.org
National Council of Teachers of Mathematics
    www.nctm.org
RadicalMath
    www.radicalmath.org
Society for Industrial and Applied Mathematics
    www.siam.org
We Use Math
    www.weusemath.org
Wolfram MathWorld
    www.mathworld.wolfram.com

Index

Text and page numbers in **boldface** refer to main topics.

abacus, 1, 83
Abbott, Edwin
    Flatland, 104
Abel, Niels Henrik, 5, 109
Achilles paradox, 32
actuarial science, 44, 45
addition and subtraction, 1–3
Adrain, Robert, 80
Aigner, Martin
    Proofs From THE BOOK, 50
air traffic control, 18
al-Banna, al-Marrakushi ibn, 99
Albert, Leone Battista, 11
al-Biruni, Abu Arrayhan, 11, 23, 83
Aleksandrov, Pavel, 125
al-Farisi, 99
Alfonso X of Castille
    Libro de los Juegos, 27
algebra and algebra education
    al-Khwarizmi and, 83
    theoretical mathematics and, 52, 55
algebraic number theory, 86
algorithms
    for multiplication and division, 76
    Groups, Algorithms, Programming (GAP), 100
Alhambra Palace (Spain), 104
Alhazen, 5, 9
al-Kashi, Jamshid, 92
al-Khwarizmi, 11
    On the Calculation with Hindu Numerals, 83
al-Kindi, Abu Yusuf Ya'qub ibn Ishaq
    On the Use of the Indian Numerals, 83

Almagest, The (Ptolemy), 134
al-Maghribi, Samu'il (al-Samaw'al), 5
al-Mahani, 92
al-Samaw'al, 5
al-Uqlidisi, Abu'l-Hasan, 83
A Mathematician's Apology (Hardy), 47
American Institute of Physics, 40
American Sociological Association Section for Mathematical Sociology, 40
American Statistical Association (ASA), 122
American system of units of length, 138
Amsler, Jacob, 18
Amsler planimeters, 18
Analyst, The (Berkeley), 145
analytic number theory, 85
Apollonius of Perga, 15
    Conic Sections, 7
Apollo's altar, 13
Approximatio ad summam terminorum binomii (a+b)n in seriem expansi (de Moivre), 80
Archimedean solids, 106
Archimedes
    circle measurement and, 32, 97
    contributions, 11, 18
    "Eureka", 70, 141
    first law of exponents and, 19
    King Hiero II crown and, 141
    simple machines and, 99
    volume of a sphere and, 13, 70
Archimedes' screw/spiral, 11, 18
architecture
    trigonometry and, 135
arcs and curves table, 134

Aristotle, 32
Arithmetica Integra (Stifel), 5
Arithmetica Logarithmica (Briggs), 20
Ars Conjectandi (Bernoulli), 99, 111
Ars Magna (Cardano), 88
Artis analyticae praxis (Harriot), 100
Aryabhata the Elder, 13, 98
Aryabhatiya (Aryabhata), 13
astronomy
    measurements in, 133, 138
atomic measurements, 138
A Treatise on Man and the Development of His Faculties (Quetelet), 72
Austin, David, 49
axiomatic systems, 3–4, 35, 36
axiom of choice (AC), 36

Babylonian mathematics
    applied mathematics, 43
    base-60 system, 91
    BM 85200, 13
    measurement, 64
Bacon, Roger, 58
Baire, René, 31
Banach, Stefan, 6, 36, 143
Banach-Tarski paradox, 36
Barrow, Isaac, 24, 34
base-2 system, 2
base-10 system, 1, 81, 83, 93
base-60 system, 82, 83, 91
Bayesian decision theory, 111, 121
"Behold!" Proof, 115, 117
Bell Curve, The (Herrnstein and Murray), 81
Bell, Robert J. T., 12

Berkeley, George
    The Analyst, 145
Bernoulli, Daniel
    trigonometric series and, 26
    vibrating string problem and, 26
Bernoulli, Jacob (Jacques, or James)
    Ars Conjectandi, 99, 111
    binomials and, 5
    curves and, 16
    exponentials/logarithms and, 20
    Law of Large Numbers and, 79, 111
    normal distribution and, 79
    permutations and, 99
    polar coordinate system and, 18
    theoretical mathematics and, 55
Bernoulli, Johann (John, or Jean)
    equation of the catenary and, 16
    functions and, 26
    theoretical mathematics and, 55
Bezier curves, 109
Bezier, Pierre, 109
binary states, 2
binomial theorem, 5–6
biomathematics, 46
Biometrika (journal), 80
biostatistics, 46
BM 85200 (Babylonian text), 13
Body Mass Index, 124
Bohr Model of the Atom, 138
Bohr radius, 138
Bolyai, Janos, 4, 96
Bolzano, Bernard, 32
Bombelli, Rafael, 14
    L'Algebra, 88
Boolean algebra (Boolean logic), 56
Boole, George
    theoretical mathematics and, 56
Borel, Émile, 31, 119
Box, George, 73
Box-Jenkins models, 73
Brahmagupta
    Brahmasphutasiddanta (The Opening of the Universe), 126
    cyclic quadrilaterals and, 97
    Hindu-Arabic numerals and, 76
    series and, 126
    zeros and, 145
Brahmi numerals, 82
Briggs, Henry
    Arithmetica Logarithmica, 20
Brun, Viggo, 86

Buckingham, Edgar, 129
Buffon, Georges, 102
Buston, Dr., 12
Buteo, Johann (Jean Borell)
    Logistica, 99
Buxton, Dr., 28

calculus and calculus education
    contradictions of infinitesimals and, 31
    history of, 22
calculus in society
    measurements and, 69
Calculus of Extension, The (Grassmann), 142
calendars
    Mayan, 144
    zeros and, 144
Canon of Samos, 15
Cantor, Georg, 30, 56
Cardano, Gerolamo
    Ars Magna, 88
    cubic equations and, 108
Cardano, Girolamo
    cubic equations and, 14
Cardan(o)-Tartaglia formula, 14
careers
    in actuarial science, 44
    applied mathematics and, 42, 44
Cartesian plane, 10, 18, 124
Cauchy, Augustin Louis
    complex analysis and, 89
    convergence and, 69
    Cours d'analyse, 32
    Green's theorem and, 143
    permutations and, 100
    probability and, 112
Cauchy-Riemann equations, 89
Cauchy's theorem, 100
Cavalieri, Bonaventura
    Archimedean spiral and, 18
    criticism of, 71
    Geometria indivisilibus continuorum, 18
    spiral curves and, 11
Cayley, Arthur, 101, 132
Central Limit theorem (CLT), 112
chaos theories, 113
charts, 28
Chebyshev, Pafnuty, 112
Chinese remainder theorem, 85

Chrystal, George, 28
circles, 32
Clarke, Somers, 27
classic mathematical problems
    double bubble problem, 71
    duplication of the cube, 8, 13
    heat conduction problem, 26
    vibrating string problem, 26
Classification theorem for Finite Simple Groups, 37
Clavius, Christopher, 113
Clifford torus (or donut), 132
Clifford, William Kingdon, 132
clock arithmetic, 85
clocks, 73
Coasta, Celso, 132
Cochran, William, 123
Codex Vigilanus (compilation), 83
coding and encryption
    Fermat's little theorem and, 6
    public-key cryptosystems, 17, 109
coffeehouses, 40
Columbus, Christopher, 134
Commercial and Political Atlas, The (Playfair), 28
commutative property, 1
compass, 3
complex analysis, 89
computational aids, 83
Computed Axial Tomography (CAT) scans, 135
computer graphics, 109
computers
    applied mathematics and, 43, 46
    theoretical mathematics and, 56
    zeros and, 145
conic sections, 7–10
Conic Sections (Apollonius), 9
convergence, 69
Coolidge, Julian Lowell, 18
coordinate geometry, 10–12, 92
Copernicus, Nicolaus, 110
Coulomb, Charles-Augustin de, 101
Coulomb's law, 101
Cournot, Antoine, 112
Cours d'analyse (Cauchy), 32
Cox, David, 56
Coxeter, Harold Scott MacDonald "Donald", 104, 105
Cox, Richard, 113
Cox's theorem, 113

Crome, Augustus, 124
crossmultiplication, 76
cryptography, 85, 98
cubes and cube roots, 12–15
cubic equations, 108
curriculum, K-12
    number and operations, 81
curves, 15–17, 68, 135

d'Alembert, Jean le Rond, 26, 32
Darboux, Jean-Gaston, 12, 143
David, Florence Nightingale, 121
De Algebra Tractatus (Wallace), 5
decimal system, 93
Dedekind cuts, 93
Dedekind, Richard, 92, 93
deductive logic, 28
De latitudinibus formarum (attrib. d'Oresme), 27
Delaunay, Charles-Eugene, 126
del Ferro, Scipione, 14, 108
Democritus of Abdera, 70
de Moivre, Abraham, 5
    Approximatio ad summam terminorum binomii (a+b)n in seriem expansi, 80
    Doctrine of Chances, 111
de Moivre-Laplace theorem, 80
de Morgan, Augustus, 101
De Ratiociniis in Ludo Aleae (Huygens), 111
DeRose, Tony, 132
Desargues, Gerard, 9
Descartes, René
    Cartesian plane and, 10, 18, 124
    conic sections and, 9
    Discourse on the Method, 11
    imaginary numbers and, 87, 130
    notation systems and, 108
    polyhedral formula and, 106
    spiral curves and, 18
Devanagari numerals, 82
De Ventula (anon), 110
Dicaearchus of Messana, 11
Diocles of Carystus, 15
direct proofs, 114
Dirichlet, Lejeune, 26
Discourse on the Method (Descartes), 11
Discovering Geometry: An Investigative Approach (Serra), 103
Disquisitiones Arithmeticae (Gauss), 83, 130

dissections, 117
divergence theorem, 143
DNA
    mapping and sequencing, 46
    topology, 53
Doctrine of Chances (de Moivre), 111
dodecahedron, 105
domains, 89
Domesday Book, 137
d'Oresme, Nicole
    De latitudinibus formarum (attrib.), 27
double bubble problem, 71
Douglas, Jess, 132
Dryden, John, 42
Dunham, William, 49
duplication of the cube problem, 8
Dürer, Albrecht, 106

earth, measurement of, 98
Edward I (King), 66
Egorov, Dmitry, 31
Egyptian mathematics, 43, 62, 81
Einstein, Albert
    elegant proofs and, 51
    field equations, 101
    Riemannian geometry and, 34
Eisenstein, Gotthold, 20
Elementary Mathematics from an Advanced Standpoint (Klein), 55
Elements
    parallel postulate and, 94
    solids and, 105
    terminology, 4
ellipses, 9
Ellis, Robert, 112
Engelbach, Reginald, 27
Enigma code, 98
Enneper, Alfred, 132
equations, polar, 18–19
equiangular spiral, 97, 98
Eratosthenes of Cyrene, 97
Erdös, Paul, 40
Erlangen Program, 132
Escher, M.C.
    Regular Division of the Plane with Asymmetric Congruent Polygons, 100
    tessellations and, 104
Euclidean geometry, 4
Euclid of Alexandria
    binomial theorem and, 5

    postulates of, 94, 95
    theoretical mathematics and, 55
Eudoxus of Cnidus, 70, 92
Euler characteristic, 132
Euler, Leonhard
    exponential functions and, 134
    functions and, 19
    geodesics and, 12
    Introductio in analysin infinitorium, 26
    Pi and, 102
    polyhedra and, 106
    seven bridges of Konigsberg and, 27
"Eureka" (Archimedes), 70, 141
exponentials and logarithms, 19–21, 134
Extracts (Stobaeus), 58

factorials, 100
Fechner, Gustav, 72
Fedorov, E.S. "Yevgraf", 104
Fermat, Pierre de
    conic sections and, 9
    coordinate geometry and, 11
    Fermat's little theorem, 6
    integral formula of arc length, 98
    probability and, 6, 119
    squares/square roots and, 130
Fermat primes, 130
Fermat's Last Theorem, 17, 86, 118
Ferrari, Lodovico, 14, 109
Fibonacci, Leonardo, 145
    Liber Abaci, 125, 145
Fibonacci sequence, 125
Fior, Antonio Maria, 14
Fisher, Ronald, 112, 120
Flatland (Abbott), 104
Flatland the Movie (movie), 104
folded normal, 81
Four-Color Theorem, 37, 50, 116
Fourier, Jean Baptiste Joseph
    heat conduction and, 26
    squares/square roots and, 129
Fourier series, 26
Fourier's heat equation, 26
Fraenkel, Abraham, 36
Francesca, Piero della, 11
Frechet, Maurice, 69
Frenet, Jean Frédéric, 11, 17
Frenet-Serret Formulas, 11, 17
frequentist approach, 111
"From Fish to Infinity" (Strogatz), 49

Fujiwara, Masahiko
   An Introduction to the World's Most Elegant Mathematics, 51
fullerenes, 105
function rate of change, 21–24
functions, 24–26
Fundamental theorem of Algebra, 88, 108
Fürer algorithm, 78
Futurama (television show), 118
fuzzy logic/sets, 113, 121

Gabriel's Horn, 31
Galileo (Galileo Galilei)
   infinite sets and, 30
   normal distribution and, 79
   proofs and, 113
   square/square roots and, 130
Gallup, George, 122
Gallup polls, 122
Galois, Evariste, 109
Galton distribution, 80
Galton, Francis, 112, 124
   Regression Towards Mediocrity in Hereditary Stature, 124
Garfield, James, 117
gas volume, 142
Gauss-Bonnet theorem, 131
Gauss, Carl F.
   arithmetic sequence and, 126
   axiomatic systems and, 4
   complex numbers and, 88
   contributions of, 12, 55
   Disquisitiones Arithmeticae, 83, 130
   linear equation simplification and, 60
   normal distribution and, 112
   polygons and, 104
   Theory of Celestial Movement, 80
   theory of curves and, 17
Gaussian curvature, 131
Gaussian distribution, 79, 112
Gaussian elimination, 60
Gauss-Jordan elimination, 60
Gauss's divergence theorem, 144
gelosia, 76
genealogy, 27
General Conference on Weights and Measures, 64
Geographia (Ptolemy), 134
Geometria indivisilibus continuorum (Cavalieri), 18
geometry and geometry education, 52, 53, 55

Gershgorin, Semyon Aranovich, 61
Giant's Causeway (Ireland), 104
Gibbs, Josiah Willard, 126, 142
Gödel, Kurt
   Incompleteness theorem, 4, 113
   ZF set theory and, 36
Goldbach, Christian, 86
Goldbach's Conjecture, 86
Goldberg, Ian, 120
Gossett, William (pseud. Student), 80
Gowers, William, 98
GPS
   trilateration and, 135
graphing calculators, 28
graph paper, 12, 28
graphs, 27–29
graph theory, 28
Grassmann, Hermann
   The Calculus of Extension, 142
Grassmann, Hermann Gunther, 12
Greek mathematics
   applied mathematics and, 43
   measurement and, 65, 71
Green, George, 71, 143
Green's theorem, 71, 143
Greenwaldian theorem, 118
Greenwald, Sarah, 119
Gregory, James, 5, 6, 18
Grienberger, Christopher, 18
Groups, Algorithms, Programming (GAP), 100
Grundlagen der Geometrie (*The Foundation of Geometry*) (Hilbert), 4
Gudermann, Christof, 11
Gunter, Edmund, 20
Gupta numerals, 82

Hahn, Hans, 143
Hamilton, William Rowan, 12, 88, 142, 143
Hankel, Hermann, 143
Hankins, Thomas, 27
Hansen, Morris, 123
Hardy, Godfrey (G.H.)
   A Mathematician's Apology, 47, 57
Harrell, Marvin, 104
Harriot, Thomas
   Artis analyticae praxis, 100
Harris Interactive, 122
Hawthorne effect, 122
heat conduction problem, 26
Heaviside, Oliver, 142

Heisenberg uncertainty principle, 101
Heisenberg, Werner, 101
Hemachandra, Acharya, 99
Henry I, King, 62
Hérigone, Pierre, 100
Hernstein, Richard
   The Bell Curve, 81
Heron of Alexandria (Hero), 97
   Metrica, 65
Hideaki Tomoyori, 102
Hilbert, David
   axiomatic systems and, 35
   contributions of, 29
Hilbert spaces, 29
Hindu-Arabic numerals
   addition/subtraction and, 1
   number and operations, 83
   place-value structures and, 76
   widespread usage, 145
   zeros and, 145
Hipparchus of Rhodes, 18, 133
Hippocrates of Chios, 8
Hobbes, Thomas, 71
holomorphic functions, 89
"How Google Finds Your Needle in the Web's Haystack" (Austin), 49
Huygens, Christiaan, 19
   De Ratiociniis in Ludo Aleae, 111
Hypsicles of Alexandria, 105

Ibn al-Haytham (Alhazen), 5, 9
Ibn Sina, 99
Identity theorem, 89
imaginary numbers, 87, 130, 135
imaging technologies, 135
Incompleteness theorem, 4, 114
Indian mathematics
   astronomy and, 98
   measurement and, 62
   negative numbers and, 145
   number systems, 83
   place-value structures and, 76, 83
   sine function and, 134
   square/cube roots and, 130
   zeros and, 145
indirect proofs, 114
infinitesimal calculus, 32
infinity, 29–31
   curves and, 17
   measurement and, 69, 71
   normal distribution and, 112

Institute for Operations Research and the Management Sciences (INFORMS), 40
intelligence quotients (IQ)
    bell curves and, 81
International Atomic Time (TAI), 74
International System of Units (SI), 140, 141
Introductio in analysin infinitorium (Euler), 26
Introduction to the World's Most Elegant Mathematics, An (Fujiwara and Ogawa), 51
Ishango bone, 75
isothermal coordinates, 12

Jenkins, Gwilym, 73
Jordan, Wilhelm, 60
Journal of Mathematical Chemistry, The, 39
Journal of Mathematical Physics, 40
Journey Through Genius (Dunham), 49

Katyayana
    Sulbasutra, 97
Kepler, Johannes
    laws of planetary motion, 9
    measurement of volume and, 65, 71
    Nova Stereometria Doliorum Vinarorum, 71
    polyhedra and, 105
Kepler-Poinsot polyhedra, 105
Kepler's Laws, 9
Keynes, John Maynard, 112
Khayyam, Omar, 5, 13
    Treatise on Demonstration of Problems of Algebra, 9
Kish, Leslie, 122
Klein bottle, 55, 131, 132
Klein, Felix
    curves and, 15
    Elementary Mathematics from an Advanced Standpoint, 55
    Erlangen program, 132
    surfaces and, 131, 132
Klugel, G. S., 95
Koch snowflake, 17, 98
Kolmogorov, A.N., 113
Kolpas, Sidney, 103
Kovalevsky, Sonja, 41
Kramp, Christian, 100
Kronecker, Leopold, 113

Lagrange, Joseph Louis, 71, 73
    Réflexions sur la résolution algébrique des équations, 100
L'Algebra (Bombelli), 88
Lambert, Johann Heinrich, 126
Lamé, Gabriel, 12
Landsberger, Henry, 122
Laplace, Pierre de, 5
    Théorie Analytique des Probaés, 112
lattice multiplication, 76, 84
Law of Large Numbers (LLN), 79, 111
laws of planetary motion, 9
Lebesgue, Henri Léon, 31, 69
Lebesgue integral, 69
Leibniz, Gottfried Wilhelm
    calculus and, 24, 26, 32
    combinations and, 99
    equation of the catenary and, 16
    functions and, 26
    limits/continuity and, 31, 32
    rate of change and, 24
    seconds pendulum and, 62
Leonardo da Vinci
    Vitruvian Man, 68
Lévy, Paul, 113
Liber Abaci (Fibonacci), 125, 145
Libro de los Juegos (Alfonso X), 27
light-years, 138
limits and continuity, 31–33, 94
linear algebra, 34
linear concepts, 33–35
linear equations, 33
linear optimization, 34
linear programming problems, 34, 39
Liouville, Joseph, 89
Liouville's theorem, 89
Littlewood, Andrew, 57
Liu Hui, 32
    Nine Chapters on the Mathematical Art, 13, 70
Lobachevsky, Nicolai Ivanovitch, 4
logarithmic spiral, 97, 98
Logistica (Buteo), 99
lognormal distribution, 80
Long Range Navigation (LORAN), 10
Lu Chao, 102

Machin's formula, 87, 88
Madow, William, 123
magnetic resonance imaging (MRI), 135
magnitudes (as line segments), 92
Mahavira, 97

Mahoney, Michael Sean, 11
Mandelbrot, Benoît, 87, 98
Mandelbrot set, 87
manifolds, 34
Maple (software), 21
Markov, Andrei, 112
Mars Climate Orbiter, 63
martingale stochastic (random) processes, 113
mathematical certainty, 35–37
mathematical modeling, 37–40
    forecasting and, 73
    global warming and, 40
    history of, 38
    for hurricanes and tornadoes, 46
    for hydrostatics, 38
    linear programming models, 34, 39
    for motion and gravity, 38
    moving fluids, 38
    for population growth, 39
    process, 37
Mathematica (software), 21
mathematician defined, 40
mathematics, applied, 42–47
    actuarial science and, 44, 45
    biomathematics, 46
    biostatistics, 46
    careers in, 44
    historical context of, 43
    for mathematical modeling, 46
    operations research (OR) and, 46
mathematics, defined, 47–49
mathematics, elegant, 49–52
mathematics, theoretical, 52–57
    algebra, 52
    analysis, 54
    geometry and topology, 52, 53
    mathematicians and, 54, 55
    number theory, 53
    training and education, 54
mathematics, utility of, 57–59
matrices, 59–61, 78
matrix multiplication, 75
matrix theory, 59
Maxwell, James, 142
    Treatise on Electricity and Magnetism, 142
mean, 72
measurements, area, 64–66
measurements, length, 66–70
measurements, volume, 70–71
measurement, systems of, 61–64

measures of center, 71–73
measuring time, 73
median, 72
Meister, A. L. F., 104
Menaechmus, 8, 11
Menalaus of Alexandria
    Sphaerica, 133
Méré, Chevalier de, 110
meromorphic function, 89
Method of Fluxions (Newton), 18
Metrica (Heron), 65
metric spaces, 69
metric system, 63, 137, 138, 140, 141
microphones, 19
military code
    Enigma code, 98
*Mirifici Logarithmorum Canonis
    Descriptio* (Napier), 84
Mises, Richard von, 112
Möbius, August, 12
Möbius band, 132
mode, 72
modular arithmetic, 85
Monge, Gaspard, 17, 28
Monte Carlo simulation, 120
Moore, Eliakim, 12
Morgan, Augustus de, 88
Moscow papyrus, 34, 64, 70
*Most Beautiful Mathematical Formulas,
    The* (Salem, Testard, Salem), 51
Mouton, Gabriel, 62
Muhammad ibn Muhammad, 107
multinomial distribution, 5
multiplication and division, 75–79
    checking results, 77
    computational speed, 78
    division algorithms, 76
    generalizing, 78
    history of algorithms for, 76
    multiplication by addition, 78
Murray, Charles
    The Bell Curve, 81

Nagari numerals, 82
Name Worshipping, 31
Napier, John, 19, 20, 134
    *Mirifici Logarithmorum Canonis
    Descriptio*, 84
Napier's bones, 84
National Institute of Standards and
    Technology, 121

Native American mathematics
    measurement and, 68
natural numbers
    series, 126
    zero and, 144
Navier, Claude-Louis, 38
Navier-Stokes equations, 38
navigation systems, 10
negative numbers, 88
    acceptance of, 13
    addition/subtraction of, 1, 2
    complex numbers and, 13
    Indian mathematics and, 145
    multiplication and, 129
Netscape Navigator Web browser, 120
Newton, Sir Isaac
    binomial theorem and, 5, 6
    coordinate systems and, 18
    imaginary numbers and, 87
    invention of calculus, 22, 134
    laws of motion/laws of gravity, 24,
        142
    limits/continuity and, 31, 32
    linear concepts and, 34
    Method of Fluxions, 18
    normal distribution and, 79
    Principia, 142
Next 50 Years, The (Stewart), 46
Neyman, Jerzy, 123
Nicholas of Cusa, 15
Nicomedes, 15
Nine Chapters on the Mathematical Art
    (Chinese text), 13, 59, 70, 128
Noether, Emmy, 143
non-Euclidean geometry, 4, 106
    parallel postulate and, 36
non-Euclidean polyhedra, 106
normal distribution, 15, 79–81, 112
Nova Stereometria Doliorum Vinarorum
    (Kepler), 71
number and operations, 81–84
    computational aids, 83
    early number systems, 81
    Hindu-Arabic numerals, 83
    Indian or Hindu numerals, 82
    Roman numerals, 82
numbers, complex, 87–90
**numbers, rational and irrational,
    90–92**
numbers, real, 92–94
number theory, 52, 53, 85–87

Ogawa, Yoko
    *An Introduction to the World's Most
    Elegant Mathematics*, 51
ogive graph, 72
On the Calculation with Hindu Numerals
    (al-Khwarizmi), 83
On the Use of the Indian Numerals (al-
    Kindi), 83
operations research (OR), 46
Opus Geometricum (Saint-Vincent), 18
Oresme, Nicole, 26
"Origin of Polar Coordinates"
    (Coolidge), 18
Oscar II of Sweden, King, 126
Oughtred, William, 20

Pacioli, Luca
    *Summa de arithmetica, Geometria,
    Proportioni et Proportionalita*,
    110
PageRank (Google), 49
Pappus of Alexandria, 97, 118
paradoxes, 29
Parallelogram Law of Vector Addition,
    143
parallel postulate, 36, 94–96
Parthenon, 15
Pascal, Blaise
    binomial theorem and, 5
    combinations and, 99
    Traité du Triangle Arithmétique, 5
Pascal's Pyramid, 6
Pascal's Simplex, 6
Pascal's Triangle, 5, 6
Peano, Giuseppe, 17, 143
Pearson, Karl, 72, 79, 124
Peirce, Charles, 32
perimeter and circumference, 96–98
**permutations and combinations,
    98–101**
Peterson, Julius, 28
philosophers
    on nature/meaning of mathemat-
    ics, 48
Pi, 89, 101–102
Picard, Charles, 89
Picard's little theorem, 89
pie chart, 28
Pierce, R. C. Jr., 20
Pitiscus, Bartholomaus, 134
    Trigometria, 134

Index

pixels, 135
placeholders, 107
place-value structures, 76
Planck length, 138
Plateau, Joseph, 132
Platonic solids, 105
Plato of Tivoli
    Academy of, 58
Plato's Academy, 58
Playfair's Postulate, 96
Playfair, William, 124
    The Commercial and Political Atlas, 28
Plimpton 322 Tablet, 118
Plücker, Julius, 11
Plutarch, 98
poetry, 41, 42
Poincaré, Jules Henri, 113
Poinsot, Louis, 105
Poisson, Simeon-Denis, 79, 112
polar coordinate system, 18
Pólya, George, 80, 101
polygons, 103–105
polyhedra, 105–106
polynomials, 107–109
Poncelet, Jean-Victor, 12, 17
population growth, 24
positive rational numbers, 91
Price, Richard, 111
Prime Number formula, 87
prime numbers, 85, 86, 130
Principia (Newton), 142
probability, 109–113
    normal distribution theory and, 112
    objective and subjective approaches, 111
    study of, 110
probability theory, 5
proof, 50, 113–116
psychophysics, 72
Ptolemy, Claudius
    The Almagest, 134
    arcs and curves table, 134
    coordinate geometry and, 11
    Geographia, 134
public key cryptosystems, 17, 109
Pythagorean School, 35
Pythagorean theorem, 117–119
    building structures and, 68
    coordinate geometry and, 11
    irrational numbers and, 91
Pythagorean Triple, 118

quadratic equations, 108, 130
quartic equations, 109
quaternions, 90, 142
Quetelet, Adolphe, 80, 112, 124
    A Treatise on Man and the Development of His Faculties, 72
Quine, Willard Van Orman, 128

Ramanujan, Srinivasa
    number theory and, 14
RAND Corporation, 120
randomness, 109, 119–121
Reed-Solomon codes, 107
Réflexsur la résolution algébrique des équations (Lagrange), 100
Regression Towards Mediocrity in Hereditary Stature (Galton), 124
Regular Division of the Plane with Asymmetric Congruent Polygons (Escher), 100
Rényi, Alfréd, 40
Rhind papyrus
    linear equations and, 33
Richard the Lion Hearted (king), 66
Richter, Charles, 19
Riemann, Bernhard
    contributions of, 89
    geometric formulations and, 34
    Green's theorem and, 143
    number theory problems and, 86
Riemann hypothesis, 86
Roberval, Gilles Personne de, 18
Robinson, Abraham, 33
Roman mathematics, 1, 82
Roman numerals, 1, 82
Romanowski, Miroslaw, 80
Roosevelt, Franklin D., 122
Rosnaugh, Linda, 104
Rota, Gian-Carlo, 42
row operations, 59
Rubik's Cube, 100
Ruffini, Paolo, 100, 109
Russian Peasant algorithm, 76

Saccheri, Girolamo, 96
Sacred Cubit, 62
Saint-Vincent, Grégoire de
    Opus Geometricum, 18
Salamis stone tablet, 1
Salem, Coralie and Lionel
    The Most Beautiful Mathematical Formulas, 51

Salmon, George, 132
sample surveys, 121–123
scatter diagram, 124
scatterplots, 123–125
Scherk, Heinrich, 132
Schönhage, Arnold, 78
Schönhage-Strassen algorithm, 78
schools
    Pythagorean School, 35
Schwarz, Hermann, 71
second (unit of time), 74
seconds pendulum, 62
Seldon, Hari, 46
sequences and series, 125–128
Serra, Michael, 103, 104
Serre, Jean-Pierre, 126
    Discovering Geometry: An Investigative Approach, 103
Serret, Joseph, 11, 17
Set Theory, 36
seven bridges of Konigsberg, 27
Shanks, William, 126
Shannon, Claude, 113, 121
Sierpinski's Triangle, 6
Sierpinski, Waclaw, 6
signal processing, 126
similarity, 128–129
Simpson, Thomas, 73
SI units, 64, 140, 141
skewness, 73
Society for Industrial and Applied Mathematics (SIAM), 44
Society for Mathematical Biology, 39
Society for Mathematical Psychology, 39
Sphaerica (Menelaus), 133
spherical geometry, 133
Sputnik, 113
squares and square roots, 129–131
Stam, Jos, 132
standardized measurements, 62
Statistical Averages (Zizek), 72
Steiner, Jakob, 97
Stevin, Simon, 92
Stewart, Ian, 46
Stifel, Michael
    Arithmetica integra, 5
Stobaeus, Joannes
    Extracts, 58
Stokes, George, 38, 143
Stokes' theorem, 143
stonemasons arithmetic, 2
Strassen, Volker, 78

Strode, Thomas, 99
Strogatz, Steven, 49
Sulbasutras, 65, 97
Summa de arithmetica, Geometria, Proportioni et Proportionalita (Pacioli), 110
surfaces, 131–133
Surveyor's formula, 104
Sylow, Peter
    Théorèmes sur les groupes de substitutions, 101
Sylow theorems, 101
Système International d'Unitès (SI), 64, 138, 140

Tait, Peter, 142
Tarski, Alfred, 36
Tartaglia, Niccolo, 14, 108
Taussky-Todd, Olga, 60
Taylor, Brook, 107
Taylor polynomials, 107
telescopes, 10
Tepping, Benjamin, 123
tessellations
    M.C. Escher and, 104
Testard, Frederic
    The Most Beautiful Mathematical Formulas, 51
Thales of Miletus, 128
Theaetetus of Athens, 105
theorema Egregium of Gauss, 131
Théorèmes sur les groupes de substitutions (Sylow), 101
Théorie Analytique des Probaés (Laplace), 112
Theory of Celestial Movement (Gauss), 80
Thomson, William (Lord Kelvin), 143
three-body problem, 126
timekeeping, 74
timescales, 73, 75
Tippet, Leonard, 120

topology, 52
Torricelli, Avangelista
    Gabriel's Horn and, 31
Torricelli, Evangelista, 98
Traité du Triangle Arithmétique (Pascal), 5
Treatise on Demonstration of Problems of Algebra (Khayyam), 9
Treatise on Electricity and Magnetism (Maxwell), 142
Treaty of the Meter, 63
Triangle Law, 143
triangulation, 135
Trigometria (Pitiscus), 134
trigonometric series, 126
trigonometry, 133–136
    functions, 89
    series, 126
Twin Prime Conjecture, 86
twistor theory, 57

units of area, 136–137
units of length, 138–139
units of mass, 139–140
units of volume, 141–142
Universal Time (UT), 74
Unreasonable Effectiveness of Mathematics in the Natural Sciences, The (Wigner), 48, 51, 57, 58
U.S. Census Bureau
    data collection, 123
    history of, 122
U.S. customary system of units of length, 138

van Heuraet, Hendrik, 98
Varignon, Pierre, 18
vectors, 142–144
Venn diagrams, 6, 28
Venn, John, 6, 28, 112
Vibrating String Problem, 26
Viète, François, 5

Vitruvian Man (da Vinci), 68
von Hippel, Paul, 73
von Koch, Helge, 17, 98

Wagner, David, 120
Wallace, John, 5
    De Algebra Tractatus, 5
Waring, Edward, 14
Waring-Goldbach, 14
Warring States period (China), 1
Weak Law of Large Numbers, 112
Weibull, (Ernest) Waloddi, 112
Weierstrass, Karl, 41, 97
Weldon, Walter, 80
Wells, Katrina, 103
Wenninger, Magnus, 106
Wessel, Caspar, 104
Westat, Inc., 123
Wiener, Norbert, 143
Wigner, Eugene
    The Unreasonable Effectiveness of Mathematics in the Natural Sciences, 48, 51, 57, 58
Wilbraham, H., 126
Wiles, Andrew, 14, 86, 118
Wilks, Samuel, 5, 80
Wishart, John, 112
Wren, Christopher, 98

Xenocrates of Chalcedon, 98
Xian, Jia, 5

Zadeh, Lotfi, 113
Zenodorus, 71, 97
Zeno of Elea, 30, 32
Zeno's Paradox, 31
Zermelo, Ernst, 36
zero, 82, 144–146
zero-point energy, 145
Zhu Shijie, 5
Zizek, Frank
    Statistical Averages, 72